The BOOK of HEAT

Stove
Spoken Here
OFFICE
VISA

The BOOK of HEAT

A Four Season Guide to Wood and Coal Heating

Edited by **William Busha & Stephen Morris**
With Contributions by the Staff of **VERMONT CASTINGS**
And Illustrations by Vance Smith

The Stephen Greene Press
Brattleboro, Vermont *Lexington, Massachusetts*

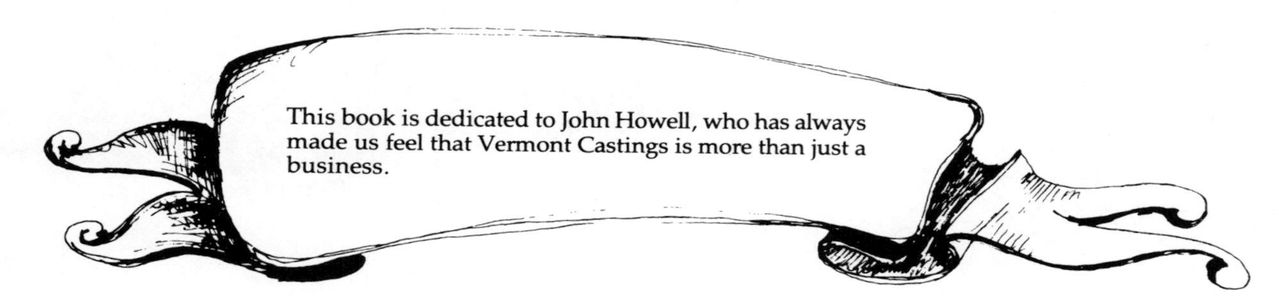

Production Credits:
Cover Design: Vance Smith
Typesetting: Neil W. Kelley
Editorial/Production Supervision: Dixie Clark
Sponsoring Editor: Robert Entwistle

First Edition

This book is manufactured in the United States of America. It is designed by Vance Smith/Dixie Clark and published by The Stephen Greene Press, Fessenden Road, Brattleboro, Vermont 05301.

Library of Congress Cataloging in Publication Data

Busha, William, 1948–
The book of heat.

Bibliography: p.
Includes index.
1. Dwellings—Heating and ventilation. 2. Stoves, Wood. 3. Stoves, Coal. I. Morris, Stephen, 1948–. II. Title.
TH7435.B79 1982 697'.22 82–12064
ISBN 0–8289–0488–X
ISBN 0–8289–0491–X (pbk.)

Contents

Preface
100,000 Tutors

Among the employees of Vermont Castings are very few with advanced degrees in combustion engineering. Although we have a Research and Development department staffed by a healthy number of certified pyromaniacs and stove nuts, they are not the authors of this book so much as the thousands of customers on whose experience we have been able to draw. A doctor on Long Island and a farmer in Nebraska may have little more in common than their experience with their Defiant stoves, but we have found that this can be a strong bond.

The need for a separate department catering to the needs of customers was first realized in 1978. Until that time the telephone caller or showroom visitor to Vermont Castings was helped by the first person who could wipe the furnace cement off his or her hands. The problem was that calls and visitors were being received so frequently that it seemed no one had time for building stoves.

Company founders Duncan Syme and Murray Howell were as reluctant to give up their roles in customer contact as they would later be to relinquish their involvement in stove construction. The decision to do so was made only after vowing that anyone who dealt with a Vermont Castings' customer would first have to pass muster as a knowledgeable stove person. He would have to know how to build the stoves and would have to have actually done it. And he would have to live with a stove in his own home, learning firsthand the joys, frustrations, and responsibilities inherent in stove ownership. On the other hand the specialist in stove building would be expected to be able at any time to drop his dead-blow mallet or grinder to come to the showroom to wait on a customer.

These basic principles have never been violated. The customer relations specialist of today is required to earn his credentials in the same way as his predecessor. Even the independent Vermont Castings' dealer from Fairbanks, Alaska, or Hilo, Hawaii is required to complete a stint of factory training in Randolph. Everyone is a *bona fide* graduate of Stove School. There is no substitute for product knowledge.

We take literally hundreds of thousands of calls from stoveowners and prospective buyers each year. We answer thousands of letters and entertain equal numbers of visitors in our showroom.

In the summer of 1981 more than 10,000 people showed up in a single weekend at our annual

Owners' Outing. These people are our tutors. Their problems are the ones that need to be solved, their suggestions evaluated, their experiences shared. At Vermont Castings we are not the teachers so much as the students. In this book we have tried to distill and to summarize our contacts with the wood and coal burners of America. It is our hope that with hundreds of thousands of tutors we can't go wrong.

The many calls and letters we receive give us a window into the world of the stoveowner. We have combined our experiences living with stoves, and are impressed by the seasonality of the subject. In the spring the talk is of next winter's fuel supply. In the summer the installation comes under scrutiny. In the fall, as we begin to operate our stoves, the talk turns to the idiosyncracies of particular stoves and operating conditions. In the winter the entire experience somehow becomes overwhelming. At once we become slaves to our stoves, and yet we find the leisure to philosophize, to tinker, and to dream.

The cycle seems as inevitable as the changing of the seasons. When the time came to choose a fundamental structure for this book, the seasonal format was the obvious first choice. From our own layman's perspective, there has been ample coverage of the woodlot by the professional forester, of combustion principles by the engineer, and of stoves through the ages by the historian. What seemed to be missing was a simple guide to the realities of life with a stove. We hope that this is what we have provided.

It has been our experience that stove ownership consists of equal parts elation, anguish, camaraderie, and monotony. On our own staff we have found that stoves provide a common meeting ground for native Vermonters and flatlanders (refugees from the populated hinterlands to the south), sheep farmers, beer brewers, actors, motorcycle mechanics, boat builders, computer programmers, and salesmen. Stoves provide a common language which people from different spheres can speak, a meeting ground for aliens who have something in common. It is as true of our Vermont Castings staff as it is of our customers.

Our book is a seasonal guide. While it has been our goal to provide useful information in this work, we have endeavored to provide hard data in a way that is fun to digest. The care and feeding of a wood or coal stove serves as a litany that assists survival during, by all accounts, the most severe season of the year. If one takes the topic too seriously, one finds the topic very dull.

This guide to heating with wood and coal can hopefully be read from cover to cover, but need not be. Our intention has been to create a guide that can be referred to periodically as well, one that will entertain and instruct, and that will enrich your own experience with your stove.

In our promotional literature we exhort our customers to give us a call, drop us a line, or come to visit us in Randolph. The common ground is stoves. The invitation seems appropriate here as well.

Acknowledgments

The creation of any book goes beyond the individual efforts of editors and illustrators. This is particularly true of *The Book of Heat*, as we relied on the collective efforts of an entire company to complete an ambitious project in record time. Our feelings can be summarized simply by saying that this work would not have been possible without the individual efforts of many Vermont Castings staff members.

Some contributors may not recognize the material they submitted in its final form, but this does not in any way minimize the importance of their efforts. Although the final product reflects a collective group effort, we would like, at least this one time, to acknowledge the individuals.

Special thanks are owed to Duncan Syme and Murray Howell, without whose original inspiration the cast of contributors would never have gathered in order to be able to create *The Book of Heat*.

Thank you all.

Charlie Page is one of the best known stove experts in the country. Although he graduated from the University of Vermont with a degree in Zoology, he had fallen in love with solid fuel stoves at an early age and had begun collecting them as a teenager.

During and after college, he worked as a logger in the Maine woods. His interest in alternative energy led first to a job with the Maine Audubon Society writing curriculum for Maine high schools, and later to work as a carpenter in the Portland area building passive solar homes.

Charlie has made several trips to Europe to research solid fuel heating there, and is an authority on European coal and wood stoves. He is currently our in-house technical resource and supervises the testing program of all Vermont Castings' products.

John King is our Warranty and Service Manager and has been with Vermont Castings for three years. He lives in nearby Braintree, Vermont, with his wife and three children. John is an enthusiastic skier, sailor, and racquetball player and cuts his own firewood. He is a coach of little league baseball and soccer, and a commissioned officer in the Vermont National Guard.

John McClain is related to Vermont Castings through marriage; his wife, Martha, is a technical writer for our R&D Department. John, a resident of Randolph, is a frequent contributor to the Vermont Castings *Owners' News* and a professional consulting forester. He is a participating volunteer consultant to the Vermont Forest Demonstration

Project, an inspecting forester for the Vermont Tree Farm System, and a member of the Society of American Foresters. John is also a member of the Vermont Natural Resources Council and the Audubon Society. He enjoys walking the Vermont woods with landowners; his frequent companion is his four-year-old daughter, Erin, who now can identify most major tree species.

Mike Smith is a wooden boat enthusiast and has studied boat building at Mystic Seaport. A well-known Central Vermont fly fishing expert, Mike spends much of his spare time chasing trout. He works in our Finance Office.

Dick Alcuri is a transplant from New York City who before joining Vermont Castings thought that the only way to get more heat into the room was to bang on the radiator. As a Vermonter, Dick has found his Vigilant coal stove to be more responsive than an apartment building superintendent. He is the authority on coal in our Customer Relations Department.

Sam Sammel spent 14 years extinguishing fires as a professional fire fighter. As a member of our Customer Relations staff, he now spends a good part of his time helping to keep fires going in the wood and coal stoves of Vermont Castings' customers. Sam is an accomplished singer and is much in demand at company gatherings. In his spare time he can be found fishing or boating.

Bebe Cameron is a Vermonter by choice rather than birth. Transplanted from New Jersey, Bebe has been an insurance auditor, a chef, and a security officer. Her talents are now put to good use as a member of our Research and Development team. Bebe operates an impromptu home for wayward animals—dogs, cats, and horses—and also enjoys gardening, fishing, and leatherwork.

Skip Peters moved north several years ago and has become permanently attached to Vermont living. Skip lives three miles off the nearest paved road in a log house. He is a veteran Customer Relations representative. When not helping customers, Skip is the host of a popular weekend show on the local radio station.

Kate Jackson acquired the taste for country living when she was growing up on a farm in New York. Her log cabin and three Siamese cats are heated with a Vigilant. Kate is a Customer Relations representative who specializes in stove problem-solving and conducting tours of our foundry facility. Before joining our staff, Kate had been a school teacher, librarian, and summer camp cook.

Doug Allen has built houses, programmed computers, and taught mathematics. Doug enjoys heating with wood and raising rabbits, and is the technical advisor to our Customer Relations staff.

Phil Neff entered the Green Mountains thirteen years ago and has been happily caught in their "mists and mellow fruitfulness" ever since. He taught school and harvested a fair amount of timber as a logger before joining Vermont Castings. Phil is part of our Customer Relations staff and works with Vermont Castings' dealers.

Jim Holmes is a graduate of Williams College. Prior to joining our Customer Relations staff he traveled extensively through Australia. Jim is a seasoned runner, and frequently runs distances some of us would consider a long drive. He is also an avid backpacker and photographer.

Steve Patterson is a native Vermonter who makes his home in the state's capital city of Montpelier. He has heated his home with wood for the last six years and as a homeowner has acquired considerable experience in alternative energy. Steve is well known as a journalist and public relations expert in Vermont and has played a key role in several state political campaigns. He is now the Communications Coordinator for Vermont Castings.

In his off hours, you are likely to find Steve behind a typewriter working on a freelance piece of writing or on the edge of a local stream after large trout.

Lynn Osborn emigrated from the flatlands of the Midwest to Vermont via Maine. Drawn to Vermont by the ideals of self-sufficiency and environmental responsibility as well as the challenge of the Vermont winter, Lynn's previous experience with woodburning as a Chicago suburbanite was limited to a few summer campfires.

Lynn is now a member of our Customer Relations staff and specializes in technical situations and answering customer letters.

Devereaux Simon came to the Customer Relations staff from Washington, D.C. and specializes in answering customer letters and coordinating special product information for customers.

Devereaux and her husband live on 20 acres north of Randolph. When not involved in restoring their old Vermont Cape she is busy skiing, jogging, gardening, and coordinating the feeding of their three wood stoves.

Wayne Staples is a native Vermonter who was born and raised on the family homestead in Northfield, Vermont. The experience of growing up on one of the few remaining subsistence farms in the state gave Wayne an early indoctrination in wood heat; the Staples' farmhouse used a wood cookstove, an antique parlor stove, a homemade double boiler, and a wood furnace.

As a problem-solver on our Customer Relations staff, he has helped many new stoveowners learn how to operate their stoves. An artist and photographer as well, Wayne enjoys helping others in his spare time. He has a second job working with retarded young adults and is a frequent volunteer for special projects involving local children or the local nursing home.

Sue Sytsma is a graduate of Vermont Technical College and has put her degree in Architectural Building and Engineering to good use by building, with her husband, their own house in Bethel, Vermont. When not pounding nails, Sue is our assistant product manager.

Brian Tyrol is a graduate of the University of Connecticut, and later the North Bennett Street School in Boston where he studied woodworking. He served as an instructor of woodworking at the school and at Harvard, and eventually operated his own furniture-making shop specializing in custom/period reproductions. Brian came to Vermont Castings as a member of our pattern shop; he now heads up our Advanced Development Group and is responsible for turning ideas for products into the real thing.

Brian is single and lives outside of Randolph with his two Huskies. He is the top-seeded company tennis player and has a small collection of rare vehicles: two Laverda motorcycles and a 1960 Rambler.

Sandy Levesque is a native Vermonter who has

has performed a variety of customer-related functions in the last few years. She has been a regular contributor to the *Owners' News* and has played a major role in organizing the annual Owners' Outing.

Sandy is well known locally as an herbalist, and has a long-standing interest both in the cultivation and uses of the many herbs she grows. Sandy is also our resident authority on stovetop cooking.

Sandy lives just south of Randolph with her husband and two children. They have spent the last few years restoring their farmhouse and furnishing it with primitive antiques.

Doris Peterson is a Vermont native and has been a member of our Customer Relations phone staff. She recently has been enrolled at Vermont College studying creative writing. Her work has been published locally and she is currently working on a full-length book.

Tim Marx is a former Customer Relations representative who is now an assistant industrial engineer in our manufacturing plant. His diverse background includes work in a steel mill, a dental equipment manufacturing plant, as a business manager of a school, and as the proprietor of his own clothing store.

Tim lives south of Randolph on the banks of the White River. His off hours are spent tending the garden, working on home construction projects, and fishing. In the winter, he is an avid cross-country skier and a speedy guard on our basketball team.

Barry DeSousa is a displaced cowboy from the Southwest who has been settled in the Vermont hills for several years. He lives with his wife and son in a house heated by a Resolute. Barry is a professional collector specializing in baseball cards, model trains, antiques, and wooden boats. He enjoys gardening, and has coordinated several local childrens' programs sponsored by Viet Nam Veterans.

Barry manages our Randolph showroom, and many of our visitors will have met and talked with him there.

Warren Needham is a native Vermonter and lives in Randolph. As a supervisor in our Customer Relations Department, Warren is well-versed in the lore of stoveland. Before Vermont Castings, Warren had been a lithograph press operator, a college counselor, and an assistant manager at a McDonald's restaurant.

Maryanne Puerner came to our Customer Relations staff with a background in biochemistry, and her science training has proven valuable in solving many of the technical stove questions asked of her.

Maryanne's summers are spent managing her large commercial vegetable garden and selling the produce at the local farmer's market. Her colleagues at Vermont Castings look to her for answers to their garden questions and stand in line to buy her superior spring seedlings.

Richard Moskwa is a member of our Customer Relations staff who homesteads on 100 acres north of Randolph. Many who have phoned Vermont Castings in the evening have received information on wood or coal burning from Richard.

At home, Richard and his wife, Maryanne Puerner, raise all their own food with more to sell. The milk from their hand-milked Jersey cow is considered the best around, and they also raise pigs and chickens.

Growing up in a farm community in northern Vermont gave *Scott Shumway* a practical view of life that he has shared with many wood and coal burners as a Customer Relations representative.

Scott served with the U.S. Army in Germany and graduated from Vermont Technical College with a degree in Agriculture. He is the high-scoring forward on our Customer Relations basketball team and lives in Randolph with his wife.

Dave Kimball is a native of Massachusetts who moved north to become the product manager of Vermont Castings. He is a product expert for the toy as well as the stove industry.

Dave is a riding enthusiast and keeps a motorcycle, a Morgan horse and a Welsh pony, and with his wife and children enjoys trail riding and hiking. He is also actively involved in programs sponsored by the local foster child agency.

Jim Holliday migrated to Vermont from his native Cleveland in 1970 because "the skiing was better." He cites a lack of critical motor skills together with a deeply ingrained desire for long life as the factors which soon led him to give up pro-skiing and take up bus driving. "But ever since I wintered-over in a 2×4-and-polyethylene-dome, my real interest has been woodstoves," says Jim, who came to Vermont Castings in 1979 as a Customer Relations representative. Since then he has become a husband, father, technical writer/illustrator, and "great deal warmer during winter."

Derik Andors is an inventor and a skilled metal craftsman specializing in blacksmithing and welding. He enjoys the seasonal pleasures of Vermont and often can be found hunting, fishing, or puttering in his garden.

Derik came to work for Vermont Castings during the company's first year of business and has been a mainstay of our Research and Development department since that time.

Bill Crossman lived in many states while growing up, from New York to California, and including Vermont. He is a graduate of Vermont's St. Michael's College with a degree in Business Administration.

When not working in the combustion lab of our Research and Development department, he can be found at his secluded home. Bill's cabin is hidden in the Vermont woods and was built by himself; he prefers to live without the modern conveniences of a telephone, electricity, or a plowed road.

Like many of our R & D staff, Bill is an inveterate inventor; he also enjoys skiing, gardening, sailing and hiking.

Bob Ferguson is the combustion engineer in our Research and Development department. He received a B.S. in Chemical Engineering from Clarkson College and spent the next 8 years working as an R & D engineer for a major corporation specializing in sophisticated alternative energy systems.

Bob lives on 60 acres located on the White River in South Royalton with his wife and two children. Much of his spare time is spent on the continuing renovation of his 150-year-old farmhouse. He still finds time for many traditional Vermont recreations: tubing on the nearby river and raising a garden and pigs in the summer, and snowshoeing and skiing in the winter.

Bob is a member of the Wood Heating Alliance assigned to the Engineering Committee, the

Efficiency Testing Subcommittee and the Heater Committee.

This book would not have been possible without the dedicated typing of the manuscript by Deb Albert and Dorian MacDonald, with additional help from Vickie Blanchard, Donna Bourassa, and Lori Grant. Kerry Smith made it possible for the project to be completed on time with her skilled assistance in researching and preparing the manuscript. Pat Crowley's guidance was essential in minimizing insults to the English language by contributors and editors, and Michael Amberger was able to get most of us together in one spot and at the same time so that he could take the group photograph.

Special thanks are in order to Laura Morris, Bill Fabian, and Jo Busha; without their unfailing tolerance, tireless support, and willingness to take on the home-chores of the editors and illustrator, this book could not have been completed.

Illustration Credits

The illustrator is indebted to the following individuals whose illustrations appeared in this book:

Barbara Carter, pp. 16, 93, 120, 154.

Jeff Danziger, p. 14. Courtesy of the Barre-Montpelier (VT) *Times Argus*.

Wendy Edelson, pp. 3, 41, 65, 75, 78, 101, 117, 123, 125, 126, 143, 168.

Ed Epstein, pp. 61, 63, 69, 100, 103, 141.

Jim Holliday, p. 108.

In addition, illustration acknowledgements are extended to the following sources:

The antique stoves on pp. 82 and 164 are redrawn from Tammis Kane Groft, *Cast With Style* (Albany, NY: Albany Institute of History and Art, 1981), pp. 32, 59, 71.

Men pouring iron on p. 83 is redrawn from a sequence by Jan Adkins which appeared in his *The Art and Ingenuity of the Woodstove* (NY: Everest House, 1978), p. 31, and in Vermont Castings' *Operational Manual for the Defiant* and *Vigilant Woodburning Stoves* in the section "How Your Stove Was Made," p. 14.

The giant power shovel on p. 105 is adapted from *Coal Facts*, published by the National Coal Association.

The cooling towers on p. 137 are adapted from a photograph by R. D. Zweig in the New Alchemy Institute, *The Journal of the New Alchemists—6* (Brattleboro, VT: The Stephen Greene Press, 1980), p. 18.

Henson's Aerial Steam Carriage on p. 155 is adapted from Michael F. Jerracu, *Incredible Flying Machines*. (NY: Exeter Books, 1980), p. 28.

The exploded view of the Pennsylvania Fireplace on p. 159 is adapted from a drawing by Robert Vogel in Paul Bortz, *Getting More Heat From Your Fireplace* (Charlotte, VT: Garden Way Publishing, 1982), p. 8.

1

Mud Season

The Spring Fling
The Amateur Lumberjack
Spring Checklist

It is Mud Season, and finally the pleasure seems to balance out the pain. Who cares if the ruts are deep enough that each trip for the Sunday paper leaves part of your car's undercoating on the road; it is heartening to see snowbanks replaced by birds returning from winter vacation. On warm days following cool nights you can almost hear the maple sap flow.

The Spring Fling

You know the day has arrived even before you open your eyes in the morning. The sounds give it away. First there are the birds, greedily and ecstatically seeking worms in ground that for the first time since October is sufficiently thawed to permit earthworm movement. Then there is the gentle drip, not of rain, but of melting snow. The last of the snow cover leaves not with a bang, but a whimper. And you can almost hear the maple sap flow.

For the first time in months you are open-eyed and alert before the alarm goes off, the bright sunshine a more gentle awakener than the clock. Instantly you know that today is not just another frustrating day in the interminable last act of winter. Today is the spiritual arrival of spring. You won't find this day specified on the calendar, but it is significant nevertheless. Today is the day that you let the stoves go out.

The unofficial first day of spring, recognized by the blood rather than the brain, is celebrated differently by many people. Some spend the day washing the accumulated winter crust from the family automobile. Others push the baby carriage around the block for the first time since fall. To one small group of people in central Vermont, this day is treated each year with a ritual celebration. The name for the event, chosen for the reckless abandon displayed to forget the rigors of a lengthy winter, is *The Spring Fling*.

To the stove builders of Vermont Castings the Spring Fling begins at noon. Following a morning of hushed anticipation unequaled since Groundhog Day, the confirming phone call from the boss is cause to lay down tools, take phones off their hooks, and cancel meetings for the rest of the day. The business of making stoves grinds to a halt in favor of an afternoon of sun worship.

The Spring Fling is a picnic. There is food, large aluminum kegs, and music. Wool jackets and heavy boots give way, in an act of faith, to T-shirts and bare feet. Visitors who arrive during the afternoon to purchase stoves often join in the fray.

One year two customers from Maine got into a poker game with some of the foundry gang and left Randolph with some spare change in addition to their new stove. There are many such Spring Fling stories.

Eventually the sun goes down, and as it is known to do in Vermont at this time of year, the temperature plummets during the night. The following morning a spring snowstorm swirls outside, and the house is 50 degrees. The wood you failed to bring in yesterday is frozen and buried. Damn!

For most, the pleasures of the Spring Fling are worth the consequences. Freedom from the fetters of stove ownership is certainly cause for celebration. But the experienced wood- and coal-burner knows that the euphoria must be short-lived or one will find himself squarely behind the eight ball for the next heating season. Mud season does not signal the end of one heating season so much as the beginning of the next. The day of delineation is the day of the Spring Fling.

The end of the spring runoff in the woods means that loggers can get their vehicles in on roads and trails dry enough for travel. The trees do not yet have their foliage, so bucking and stacking are infinitely easier. The weather is cool enough to moderate a legitimate sweat, and the bugs have not yet hatched into their pre-summer swarm. In short, this is the perfect time to be working in the woods, before Mother Nature begins her photosynthetic wonders.

Once wood is out of the forest, it needs to be processed in time to take full advantage of the solar benefits of summer. Unsplit wood has only a fraction of its surface exposed to the light and air compared to split wood, so the work done now will produce the maximum drying benefit. At Vermont Castings the wise, experienced veterans know that this is the time to sign up for the company splitter. By the time September comes around, this piece of equipment is more in demand than a dogbone in a kennel.

So the wise woodburner puts up his winter wood supply the previous spring. The true professional, however, goes a step beyond. Al-

Mud Season

Mud Season is Vermont's ugly stepchild. While Spring brings daffodils, cherry blossoms, and flowering dogwoods to the habitable portions of North America, Vermonters, with the hereditary stoicism of the hardscrabble Yankee farmer, confront the seasonal phenomenon known here as "Mud Season."

Following a long and arduous winter one would think that nature would smile beneficently on the northlands, but no. By early March the air warms just enough in the daytime to melt the uppermost layer of snow, which then refreezes at night, leaving a jagged crust to ruin the cross-country skiing. The same melt-freeze syndrome affects the roads. The paths that could be travelled solidly during winter now become a network of spongy grooves deep enough to bury any four-wheel drive vehicle to its hubcaps. As the snow run-off adds surface water to the thaw, the world becomes a soggy purgatory. Most of all, Mud Season is a state of mind. The sufferer from cabin fever now faces an outdoors which is less accessible than ever. Hopes for warmth and light are dashed by sadistic late season storms. Once the ordeal is over, the survivors joke and brag . . . and save money for a trip South next year.

though he follows the same schedule, the wood supply he is manufacturing is not for the next heating season, but rather for the one after that. If one-season dried wood is good brandy, two-season dried wood is ambrosial cognac. Somewhere, sometime in one's life a quantum leap must be made to achieve this elite status among stoveowners. Once the commitment to get a year ahead has been made, the dividends may be collected indefinitely.

The first of June arrives. The fools in your neighborhood have squandered their time putting in gardens, fishing, bicycling, canoeing, and sipping gin and tonics on the front porch. You, meanwhile wipe the sweat from your brow and stare at next winters' fuel supply, smug in your embracement of the Yankee work ethic. Each party feels certain that he has found the better way. As the season rolls onwards, and summer no longer springs eternal, the odds change increasingly in your favor.

The Amateur Lumberjack

The last few decades have seen the development of two seemingly opposite trends: The first is the rush of technology, particularly in the area of electronics and synthetic materials; the other is the desire to gain more control over one's life through the pursuit of a more basic lifestyle. The latter is essentially a reaction to the former. Like children seeking comfort in the security of parents, many of us seek reassurance, in the face of rapidly escalating technology, in a basic relationship with our environment. We grow our own food, make our own clothes, fix our own cars, and gather our own fuel.

Burning wood is an exercise in consciousness, an assertion of one's position in an ecological cycle. There is a primitive comfort in handling the chunk of wood that will provide immediate warmth; unlike gas, or oil, or electricity, the chunk of wood is tangible, something real.

Burning wood is also a commitment, one that requires some understanding of the impact of the act. New chainsaw owners often are surprised at how difficult it is to fell their first few trees. The act demands that a decision be made to terminate what often is several decades of growth in a living organism. It is a humbling experience to stand beneath a tree that has occupied the same spot on the earth for 100 years. The decision to remove the tree and burn it cannot be a casual one.

Perhaps the most gratifying aspect of burning wood is the knowledge that the informed harvesting of selected trees is a sacrificial but healthy process; the rest of the forest benefits, and the burner of the wood benefits.

It doesn't matter whether you buy your wood from someone else or harvest your own woodlot: You will want to understand your impact on the environment.

A knowledge of basic forest principles will lead to a greater philosophical peace each time you stoke the stove.

Managing for Wildlife

A poll of Vermont landowners a while back produced a surprising response. When asked to list objectives and concerns about their woodlands the group listed wildlife first, above timber, firewood, recreation, aesthetics, and all other interests. In other areas wildlife may not head the list, but it is certainly near the top. In order to manage a woodlot for wildlife, it is important to know that habitat diversity is one of the most important factors. Timber cutting in a manner to leave a variety of age classes and a mix of species contributes to habitat diversity. There are also several special practices that can be accomplished by the small woodlot owner.

As a start, consider thinning around, or "releasing" wild apple trees. In many woodlots, these trees are common. They are present because seed has been deposited by domestic and wild animals. They are important as a food source for deer, grouse, rabbits, and squirrels. As the forest matures these trees become overtopped and decline in vigor. To revitalize individual apple trees, remove all shrubs and trees that are competing with the crown. Removing dead wood and vertical suckers will encourage fruit production. If possible, fertilize the tree every few years to encourage further growth.

Aspen, at various stages of its life, provides excellent food and cover for ruffed grouse. Most aspen stands may be regenerated by clearcutting small patches. The species has the ability to reproduce by root suckering. The resulting stand is usually extremely dense if aspen were present in great enough numbers before the stand was cut. The size and nature of the cut may vary with surrounding vegetation. Usually, cutting adjacent stands in an organized pattern over a period of time promotes the best habitat for grouse. To aid in planning a habitat cut, technical assistance should be sought.

In large woodlots, proper management has an even greater impact. The white-tailed deer in northern climates, for example, seeks winter cover in dense conifer stands that have a southerly aspect. The timber present is commonly mature and hence ripe for harvest. In such a case, proceed with care and make a well planned and supervised light cut. Secure the advice of a wildlife biologist or well-informed forester. A heavy, thoughtless cut in a large deer yard could bring disaster to a local deer herd.

"Snags" are dead or dying trees that offer nesting sites to many birds. Many woodburners see this as an excellent source of fuel. Dead wood, however, is usually in the initial stages of rot. Rot lowers BTU output. Furthermore, dead trees do not compete for crown space with live ones and do not need to be removed in a thinning. Snags that are hollow or contain cavites may offer denning sites to squirrels, raccoons, songbirds and other wildlife. When you thin your woodlot, try to save at least five snags per acre.

If you would like to encourage wildlife, it is beneficial to learn as much as you can about the habitat of each species you are interested in. Good sources of information are available from state fish and game department and the U.S. Fish and Wildlife Service.

Evaluate the Woodlot

If you are fortunate enough to have a piece of land on which to cut your own wood, or if you know someone who has land and is willing to let you cut on it, you will want to know something about forest health.

You can leave your forest alone as it has probably been left for years, or you can manage it to harvest firewood, improve the growth and value of your trees, increase your awareness of the natural world around you, and enhance the scenery.

A day in the woods with a chain saw and without knowledge of what you are doing can create a disruption of the forest growth pattern that may take decades to correct. Plan to spend a considerable amount of time in your woodlot before you start the chain saw.

Doodlebugs

No discussion on the various ways to transport wood is complete without saying something about doodlebugs. You won't find one of these unique vehicles on your local truck dealer's lot, and you probably will never pass one on the road. But nevertheless, each year doodlebugs haul tons of firewood and pulpwood out of the woodlots of Vermont as well as other rural states.

Doodlebugs are homemade, off-the-road vehicles. They are made from old trucks and sometimes cars. They consist of a frame, a flat bed on which to place wood, an engine, a transmission, and wheels. Usually there is a driver's seat and sometimes there is a steering mechanism. Some have brakes. They are working junk; cutdown, welded, modified, bolted together marvels of backyard engineering, leading a charmed second life.

The tradition of doodlebugs goes back more than 40 years before so-called off-the-road vehicles became popular. All that was needed to start the trend was a supply of discarded junk automobiles and a few inventive types who couldn't bear to see a still useful item thrown away. In the early 1930's, one could buy a conversion kit from Sears Roebuck to turn a Model T car or truck into a tractor. The kit consisted of two large metal wheels and an axle. The axle was bolted onto the frame in front of the existing rear axle, and on it were mounted the large metal tires. A gear on the existing rear axle fitted into a larger gear mounted on the spokes of the metal wheels to make the tractor go.

Most doodlebugs have not been built out of manufactured kits, though, but rather out of what was at hand or could be picked up at the dump; this latter resource led to an alternative name. Doodlebugs are known in some areas as "dump picker specials."

Whatever you choose to call them, each machine is created to suit the needs of the inventor and adapted to meet the specific requirements of his land. No two are completely alike. A doodlebug may or may not be just what you need to get your wood out, but look for one anyway. They are splendid examples of native American ingenuity.

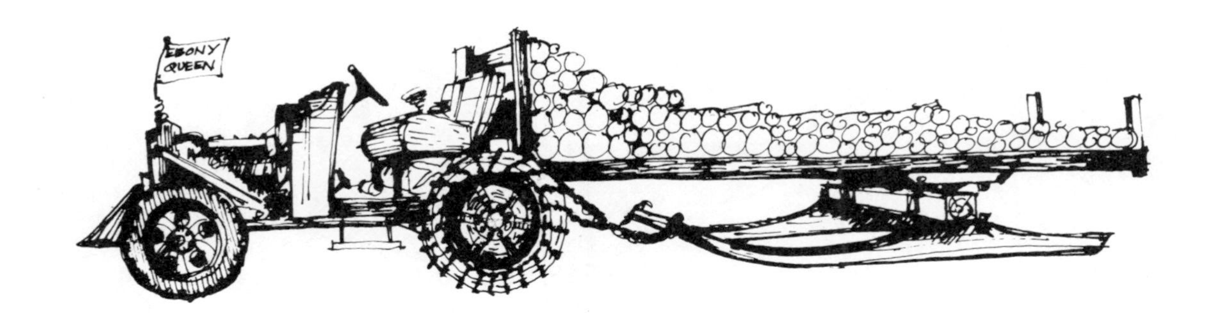

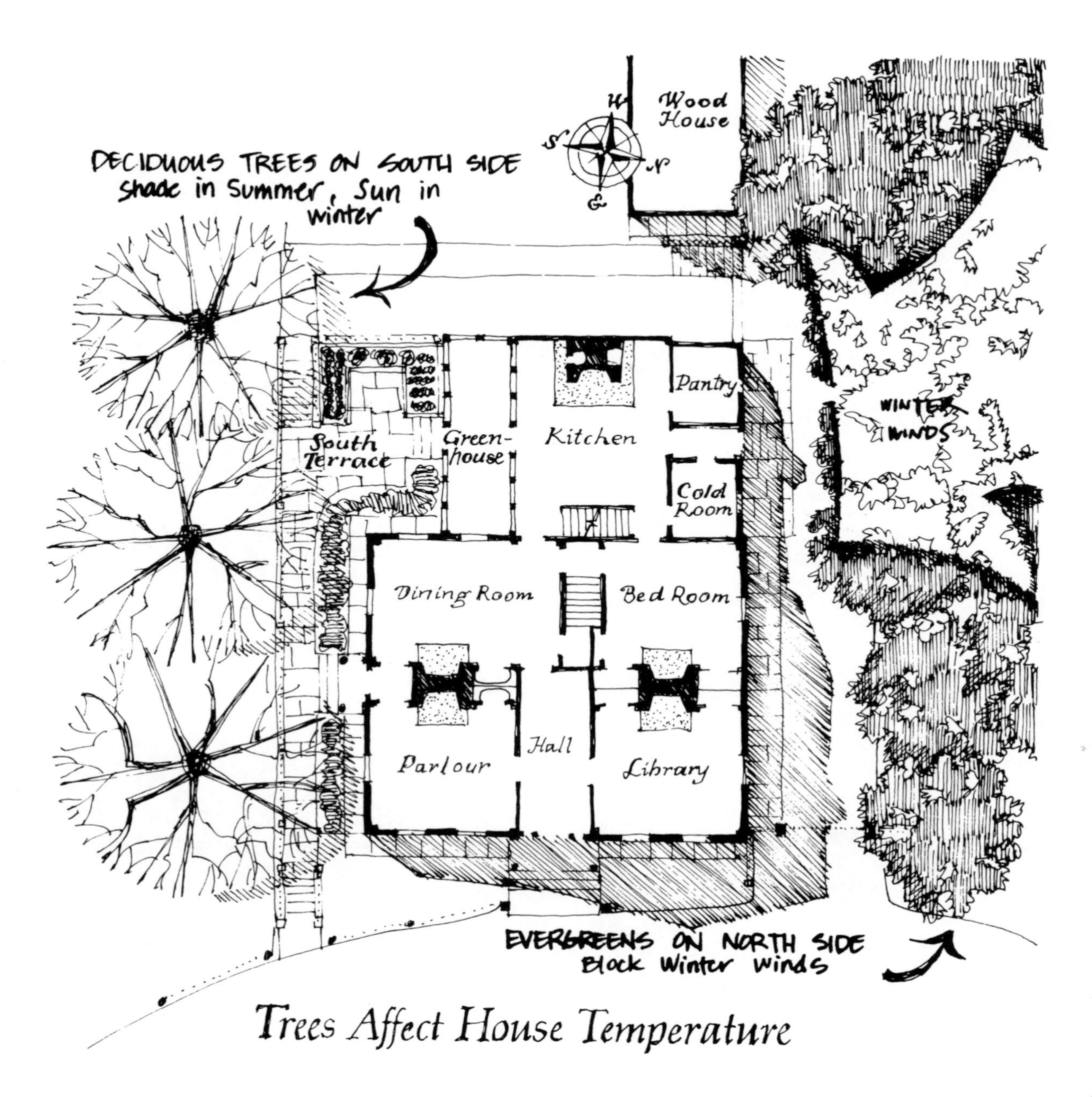

Trees Affect House Temperature

First, let's take a look at some of the objectives you may have for your woodlot. As a woodburner you may be most interested in producing firewood. But this is probably not your only goal; watching the raccoons from your porch and identifying birds out the window are also important. Others may enjoy picnicking on the rocks under an old oak tree or skiing among the firs in winter. Another consideration is the effect of shade upon the internal temperature of your home, both in summer and in winter, and how to make the trees you have fight the energy crunch with you. List all your goals and decide which are the most important.

Knowing why you are going to do something is the first step in forming a management plan. The how comes later.

Boundaries

Locating your woodlot is more difficult than it sounds. Many woodlot owners have an idea where this boundary line is, but are surprised when they follow a disappearing fence to a non-existent corner. Knowing, defining, and marking your boundaries will save you grief and money once you have started cleaning, thinning, or harvesting. Cutting someone else's timber, known as timber trespass, can be a devastating experience. The penalty for this varies by state, but can sometimes go as high as three times the value of the trees removed, plus damages and legal fees. If you do not know your boundaries you should have the property surveyed and the title searched by a registered surveyor. The results of surveys can be extraordinary. One local survey found a property to be twenty-one acres less than previously thought, and paid for itself in savings on the owner's taxes.

The Stand

Once you know what woodland really belongs to you, walk through the woods. If the trees are all the same in species, age, and condition then you have what is known as a "single stand" of timber. A stand is a group of trees that is uniform enough in character to be distinguishable from other surrounding stands. In a very small woodlot, only one stand may be present. As acreage increases, the chance of two or more stands increases. Generally speaking, the stands that collectively become a forest are managed separately.

Professionals often locate stands through the use of aerial photographs in conjunction with on-the-ground observation. These two sources will provide information for a forest-type map showing location and acreages of various stands in a property. This map can be extremely helpful in laying out roads. Also available are soil surveys and topographic maps to be used with the owner's map. Notes concerning quality and type of timber present on the property are also included.

After defining your stand, the next step is determining what condition each stand is in and what needs to be done to improve health, vigor, and quality. To do this you need to know how to evaluate these conditions. It is at this point that many owners realize the need for professional advice. If you are in doubt about the condition of your land, seek competent help. An understand-

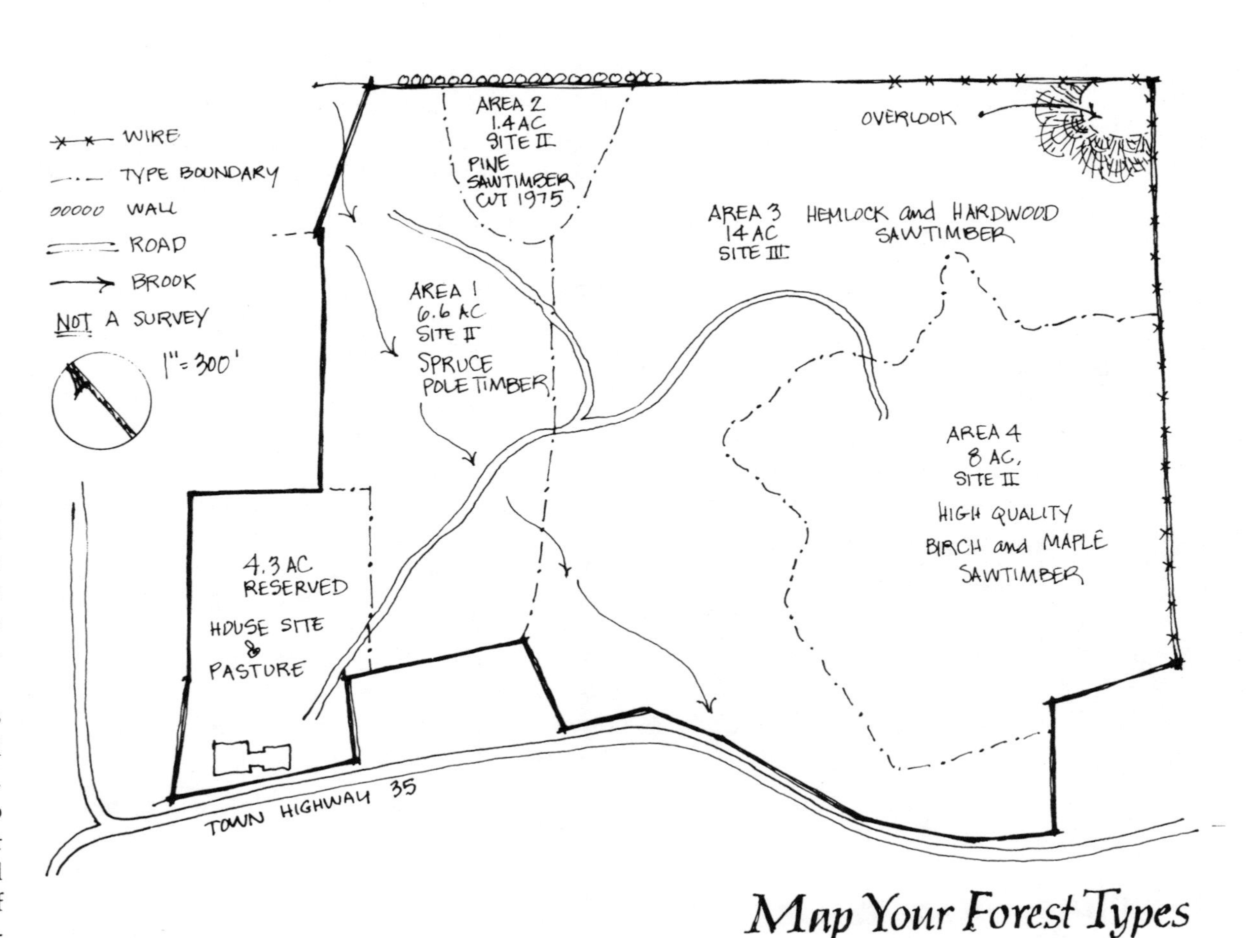

Map Your Forest Types

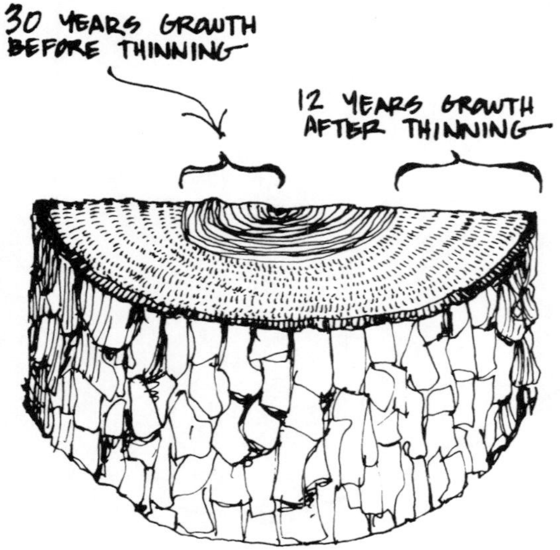

ing of some basic principles will help you to better appreciate the management recommendations that are ultimately made.

Size. First consider size. Trees may be categorized by size according to height and diameter. The forester's term of qualification is "diameter at breast height" or DBH. This is the diameter in inches of the tree at a height approximately even with the forester's breast. Since foresters often are well over or well under six feet tall, the term DBH is technically qualified as 4.5 feet from the ground.

Starting from the ground up, the progressive size of a stand begins with seedlings or sprouts. These tiny individuals are often overlooked in the woods and frequently are stepped on as you are walking. They occupy the space between the ground and four and one half feet above the ground and have diameters of less than an inch.

The next larger group are the saplings. Their DBH is from one to four inches. Over four inches, trees are known as pole timber until the diameter reaches about ten inches. They are then classed as saw timber. This final category is not rigid, however, and most saw timber harvests concentrate on trees larger than ten inches. Since volume of wood increases dramatically with a growth in diameter, it would be poor use of the resource to cut all ten-inch trees without further consideration of future growth.

Site Evaluation. How high a tree will grow depends on site potential. These are the biological and geological characteristics of a piece of land. Short of fertilizing, we can have little positive influence on the soil's capacity to provide nutrients for growth. We can, however, destroy the earth's potential by letting water flow directly down steep embankments, or cutting the woods so drastically that increased water flow takes away the nutrients and deposits them downstream.

Site quality is usually measured by comparing the height of a tree with its age. This comparison gives us a measure of potential called a site index. The site index will vary by species, and each species on a site will have a different index, expressed numerically. The higher the number, the better the site for the species.

Density. Density is one of the most important factors involved in prescribing treatment for a stand; the diameter growth of individual trees is directly dependent upon it. Diameter growth is easier to influence than height. It is something we can increase on preferred trees through thinning. Trees grow outward in direct relation to how much space is between them. In most unmanaged stands, the number of trees per acre is quite high. Trees that are close together are generally small in diameter. If we thin the stand, each tree will have more growing space and consequently a larger crown. A larger crown, or top spread of the tree, means more leaves and more photosynthesis. That is, the tree has more capacity to take in the sunlight and carbon dioxide it needs to live. This all adds up to greater diameter growth. As your trees increase in diameter they will need more crown space. Hence, more thinning will be in order on a continuing basis.

Optimum density levels vary at different stages of the growth of a stand. In a young sapling stand, it is advantageous to have a high density level to reduce low branching. Branches, when

30 YEARS GROWTH BEFORE THINNING
12 YEARS GROWTH AFTER THINNING

Thinning Increases Growth

sawn into lumber, become knots and are considered defects. As the stand grows in height and diameter, a less dense situation favors increased diameter growth. Pole-sized hardwood stands produce excellent firewood when thinned.

Quality. Trees of high quality have long straight stems and are void of branches, knots, or branch stubs for at least the first sixteen feet of the tree. They should also be free of crooks and sweeping curves, and they should stand erect rather than lean.

Health. The general health of a stand is reflected visually in many ways. Foliage diseases are usually extremely apparent. Look for early turning or skeletonizing of the leaves, or browning of parts of a crown. Stem diseases and maladies of the internal structure are frequently noted by the presence of fungi growing on the bark. These fungi are fruiting bodies of a more serious problem under the bark where moisture and nutrients are being consumed as the wood rots.

A Management Approach

As a starting point, look at an established dense stand of pole timber. You will notice that some trees are taller than others. This is true in stands of both deciduous trees and in softwoods. The division of these trees of varying heights into crown classes is an important step. Four crown classes are commonly recognized. The crowns of many trees that stand together in a forest form the main "canopy." These trees that form the main canopy are known as "co-dominant" trees. The more vigorous trees in the forest have crowns that extend above the main canopy, and these are called "dominant" trees.

Trees that do not completely reach into the main canopy are known as "intermediates." The weakest and shortest stems that do not get any light through the canopy are "suppressed" or "overtopped." In most cases, these trees will not live. Looking over the canopy, note the trees that fall into the dominant and co-dominant classes.

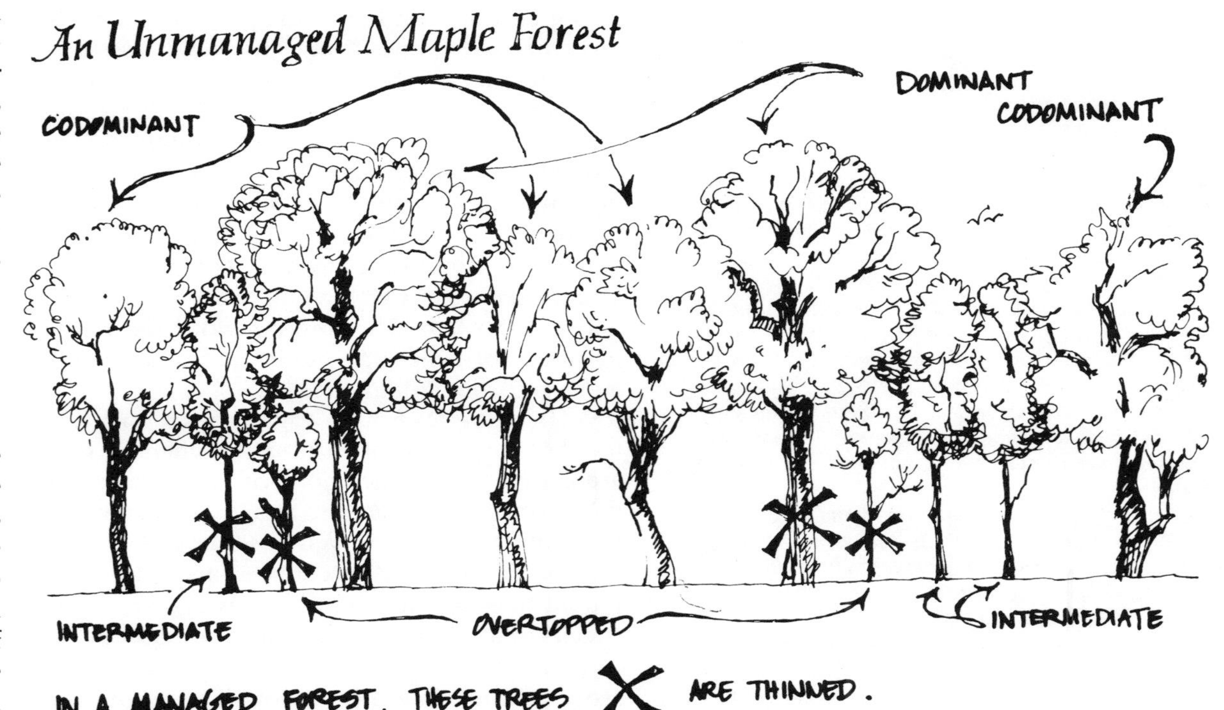

A Solid Investment

A lifetime of forest ownership can be a rewarding experience. Trees, like savings accounts, grow. Unlike savings accounts the woodlot may be used before, during, and after maturation.

On selected high quality sites in Vermont the rate of growth in value has been as high as 16% in real dollars—above and beyond inflation! The rate, unfortunately, does not hold true for all woodlands, but all trees do grow. The expected growth in stumpage rates parallels that of demand: a 100% increase, again in real dollars, is expected by the year 2000. The increasing real value of timberland makes the investment in management all the more worthwhile.

For tax purposes, the costs of management including surveying, hiring a forester, road construction and legal fees may be appreciated over the length of ownership. Income derived from a timber sale is treated as capital gains and about half the total gain is taxable income.

Other economic considerations include Federal cost sharing programs administered by the Agricultural Stabilization and Conservation Service (A.S.C.S.). These programs help the landowner afford to make improvement he otherwise could not afford. Check with your local office to find out what programs are available in your area, and how to apply.

In many states owners of forest land are given an incentive to participate in the management process in the form of a property tax break. The reduction in taxes usually requires proof of management in the form of a written management plan and map. In some states Tree Farm Status is sufficient proof. Check with your state department of taxes or your local tax assessors to determine what is necessary for your woodlands.

These are usually the most vigorous, and if of good quality should benefit from a thinning.

There are other factors involved in planning a thinning. One is "species differentiation." If you have done your homework or engaged a forester, you will know which species or group has the most potential on your soil and can thin accordingly. In the case of tree types of differing values, favor the species of greater economic worth. Another type of cutting during a thinning involves the removal of some of the dominant trees. In some woodlots, dominants have gone past their prime and are hindering the growth of a more vigorous stand underneath. In certain cases it is better to kill these trees by girdling rather than to try felling them and risk injuring other more favorable stems in the process. Girdling may be accomplished by cutting deeply through the bark in a circle around the tree. The cut must be continuous, as if you were cutting along a piece of twine tied around the circumference of the tree. You must cut through the inner bark and into the meat of the tree. This process stops the transportation of nutrients. Girdled dominants will make excellent wildlife trees.

There are four main approaches to management of your woodlot by cutting. Generally, they are known as "regeneration" cuttings, and as the name implies, the goal is to stimulate new and continuing growth as well as to harvest timber. The first, called the "selection" system, cuts individual trees as they mature. This system favors species that do not require much sunlight to become established. The holes in the forest canopy are small and allow only minimal light to enter the stand. To the untrained eye this may be the most appealing type of cut as it causes the least visual disturbance. There are some problems associated with the system, however. The first is workability: how to get the tree down and out without scarring too many other stems. The second is taking only the biggest and best trees out of the forest. This is not such a bad thing if the land has been managed in the past. If it has not, selectively cutting only the best can lower the overall quality of the stand. This situation is known as "high grading." Repeated high grading has been responsible for the deterioration of thousands of acres of woodlands. Always cut the worst first.

A second system of reproducing a stand of timber is the "shelterwood" method. This method involves several cuttings. The first cutting is a preparatory cut and removes low value trees. This allows the better trees to enlarge their crowns, and stimulates seed production. The second cut is the "seed cut," which gives more light to the forest floor. The remaining, best trees are responsible for seed production and establishing the new young stand. The next cut(s) are called removals in which the overstory of seed trees is removed, giving the regeneration tree room to grow. Advantages of this method are that reproduction is fairly certain; the resulting stand is protected, and the seed comes from the pick of the best trees. The disadvantages include aesthetics (due to the heavy seed cut necessary) and the low financial value of the initial cuts.

"Clearcutting" is perhaps the most widely disputed and controversial practice in forestry today. In its proper sense, it literally removes every tree in a given area. Seeds from adjacent stands are depended on to restock the cut area.

The seed tree approach is related to clearcutting but retains certain trees to supply seed for regeneration, hence the name. The seed trees are small in number and should be well spread out over the cut area to ensure good distribution of the seed. In both of the above methods, visual impact is tremendous. Recently clearcut areas appear devastated. Also, protection against erosion may be diminished. If natural regeneration fails using this method, the only way to restock the site is through planting. On the plus side, clearcutting is an easy method of removal. There are no residual trees to be injured, and larger more efficient equipment can be used.

Any of the above methods may be used to regenerate a timber stand. Which method to use depends on the condition of the stand prior to harvest and the nature of the species that you are trying to establish, as well as the soil, steepness, and erosion potential of the site. Each method is applicable on areas from five acres on up in size. In each system there is room for mismanagement of the resource. A poorly designed selection cut can conceivably be more detrimental than a well-planned small clearcut. The system you employ should be decided on after careful consideration and professional advice. It is wise to remember that the tree you cut today may not be replaced in size and stature for several generations.

Professional Forestry Services

Many woodland owners are not aware of sources of help in planning and conducting their management activities. There is help available. In some cases it is even free, as with a county forester.

The role of the county forester is slightly different from state to state, but is generally one of an introductory nature. Because of the wide range and number of residents to be served, the time a county forester can spend on each person's woodland is often limited, either by policy or by the sheer size of the workload. In some states county foresters provide the service of disseminating information and recommending further sources of aid in managing the home woodlot. As an example, Vermont county foresters may spend one day or more with an owner marking firewood, planning, and making general recommendations about the woodland. In the case of timber sales, the county forester may mark the trees to be cut, prepare the contract, and contact a logger. As an alternative, the forester may refer the owner to a list of qualified consultant foresters. The owner then chooses one and works with that forester on a management plan.

Consultant foresters are qualified professionals who, in some states, are licensed or registered by the state(s) in which they work. Their responsibilities are the best interests of their clients as well as sound silvicultural practices. For their services they charge a fee. In the case of timber sales, a percent of the gross income from the sale is usually levied. This fee varies by individual and by the job, often depending on the type of cut, amount of work necessary, and frequency of supervisory visits to be made while the timber is being harvested. Other services performed by consultant foresters may include laying out roads, marking boundary, determining

Caveat Emptor

Last year a neighbor complained that his wood supply was dwindling much faster than seemed appropriate for the mild fall weather. We decided to solve the mystery by measuring up the cord wood and not burning any of it for awhile. We also surreptitiously notched a few randomly selected logs in the pile and dabbed them with red paint. Our suspicion that someone was helping himself to the woodpile was confirmed a few weeks later when the neatly stacked wood was reduced by a quarter of a cord.

The case of the missing wood was solved a month later when a load of wood I had ordered arrived. Several of the logs bore our notched brand and were artfully decorated with red paint. I happened to mention this unusual phenomenon to our wood dealer, and he too marveled at the coincidence, though a few weeks later a full cord of unordered wood was deposited at my house and that of my neighbor's.

Since I try to look at the bright side of any unfortunate development, I decided to accept the additional wood without any further comment. That's not to say I didn't recognize the moral of the experience: As with the purchase of any commodity, caution is always advised.

Here are some practical suggestions if you're setting out on your first woodburning experience:

Talk to friends and neighbors who heat with wood. Are they happy with their dealer? Does he deliver on time? Are his prices competitive? Does he deliver the type of wood promised?

Ask the dealer questions. Will he have enough wood to help you out if you miscalculate your needs? Can you get a price break by ordering multiple cords? When was the wood cut? (A particularly important question if you're ordering seasoned wood.)

If you're unsure, try ordering a single cord from two or three different dealers. Also, call your state energy or forestry office. They may have a list of dealers.

Most importantly, be specific when you order. Don't hesitate to ask for specific length of wood and measure it upon arrival. Nothing is more frustrating than trying to fit a 25-inch piece of wood into a stove that will only take a two-foot chunk.

wildlife habitat potential, marking trees for firewood, forming management plans, creating a forest inventory, and generally providing advice to the landowner about all phases of forest management.

A third resource for the woodland owner is the United States Department of Agriculture Soil Conservation Service (S.C.S.). On areas where soil mapping has been done, the SCS may be able to provide an owner with information about the potential of the land under the timber for site quality and tree growth, specific advice on what trees to plant if necessary, and information on water resources, pond layout, road layout, and possible sites for sources of gravel to be used in road building.

A final resource is the Extension Service of your state university. The Extension Service is primarily responsible for providing information and educational materials. There are many good publications in pamphlet or short book form that are available from them. The scope of the Extension Service is very wide, taking in virtually every aspect of agriculture including fruit tree care and woodland information.

Obtaining Firewood

Think of your woodstove as a child or a pet. It depends on you, a commitment is made by keeping one, it is a lot of work; at the same time, it returns a great deal of pleasure. The wood stove can be a key element in shaping your lifestyle, and you can invest as much or as little time in it as your choose. The greatest amount of time will be spent if you start with standing trees as described in the previous pages on woodlots. The least amount of time will be spent if you only stack the wood and feed the stove, and pay someone else to do the cutting and splitting. Almost all of the work is enjoyable though, and you only have to do it once a year. (Once a year for four months, as the joke goes.)

Working up the wood supply is therapy as well as recreation. In a physical sense it is good exercise. Six cords of maple that is first stacked outdoors to dry and later moved to the basement for use requires that you move around 50,000 pounds. You could pay a fair amount of money to your local health club for the use of their weight lifting equipment for a similar amount of exercise, and then have to pay fuel bills on top of that.

Splitting wood is a good form of emotional therapy as well. Many of us like to split a few chunks after a hard day at the office, finding a tacit satisfaction in being able to strike something with unbridled force. This is a great way to release tension and a lot better for you than a stop at the local bar and grill. After-work splitting or stacking wood gives enough pleasure to be habit-forming, and you may end up wondering what your oil-burning neighbor does in the evening to relax. Consider this scenario for your oil-burning counterpart:

"Well Mary, it was a hard day at the office. I think I'll just pick up the phone and call the oil man. Maybe he can deliver a couple hundred gallons."

Enough of the theoretical. Let's consider next year's wood supply.

A Good Supplier

Buying firewood is more of an art than buying conventional fuels. Wood is a natural material with all the individuality and beauty of living things. Every piece of wood is different, and one measure of your skill as a woodburner is your ability to judge what you are getting when you pay for a load of firewood.

How much wood are you getting? Firewood is usually sold by volume, and the standard measure is the cord: 8 feet by 4 feet by 4 feet. In many areas wood is sold by the "face cord" or "short cord," a measure 4 feet high by 8 feet long, but only as deep as a single log, be it 12 inches or 24 inches. There's nothing wrong with buying by the face cord; just find out how long the logs are, calculate the fraction of a full cord that you are getting, and expect the price to drop proportionately.

A full cord contains 128 cubic feet, but because there are air spaces in it the solid volume of wood can be anywhere from 65 to 100 cubic feet. Arguments abound as to whether you get more wood with large logs or small, split or unsplit; but most people agree that a mixture of large logs, with small ones to go in the spaces, gives you the most wood per cord.

Whether you buy by the cord or the face cord, you should get your measure in fairly tightly packed logs. You may want to stack it more loosely for drying, but be sure you are not paying for big air spaces.

There is a variety of other measures of firewood, none of them precise. A truckload depends on the size of the truck. A run means a truckload. A rick simply means a stack, and the size varies from dealer to dealer. Try to get the price in terms of cords or cubic feet so you can compare. In one case we heard of, three full cords of wood were ordered from three different dealers. The specified length had been 16"–18", cut, split, and delivered. All of the wood was understood to be unseasoned.

One dealer delivered a stack measuring just over 100 cubic feet (counting air space) for $55. Another came in with 126 cubic feet at $65. And the third came in with 130 cubic feet for $70. The differences in wood delivered by two of the dealers was miniscule. The final decision was to buy the rest of the supply from the $70 a cord dealer because none of the logs exceeded the specified 18" maximum and none of the load needed to be split again to fit the stove, a critical consideration if you would rather spend your time trout fishing than swinging an axe.

What services are you paying for? You can pick up your firewood yourself, have it thrown on your lawn in a heap, or have it neatly stacked in your shed. The more work the seller has to do, the higher the price will be. Make sure you know the terms when you agree on a price.

What size are the logs? You can save considerably by purchasing log length firewood and doing all of the cutting and splitting yourself. But before you do this, picture a heap of twelve-foot tree trunks maybe two feet in diameter, and imagine yourself attacking them with a chain saw. Many of us find this sort of work rewarding: it's good exercise, and you'll end up with firewood cut just the way you want it. But it is a lot of work.

Some dealers sell 4-foot logs. These also need cutting and splitting, but the job isn't quite so overwhelming.

You can buy the whole logs cut to stove lengths, and just do the splitting yourself. Be sure to get the right length for your stove. If your wood is too short you won't get as long a burn as your stove is capable of. If your wood is too long you'll be a very unhappy woodburner next winter.

Probably the most popular method is to buy full cords of wood "cut, split, and delivered." Prices for this service will vary dramatically depending on the locale in which you live and how long the wood has been seasoned. The price of fully processed and delivered wood will vary greatly. Obviously, the closer you live to the supply, the lower will be the price. However, city dwellers may pay twice as much for firewood as their country cousins and still find it less expensive than paying for electricity. Compare prices. Some dealers are finding ways to process wood more inexpensively than in the past to give them a competitive edge in the market.

Wood Quality

Dry or green? A big advantage of buying in the spring is that you can get green wood and season it yourself over the summer. There is usually a considerable price difference between green and dry wood. If you do buy dry wood, make sure that it is really dry. Signs of seasoned wood are splits on the ends of the logs, relatively light weight, and the sound of two logs struck together. (Green wood thuds, dry sounds more hollow.)

What kind of wood? If you have a choice, get the heaviest wood you can. Any given weight of wood contains about the same heat potential (which is measured in BTU's), whatever the

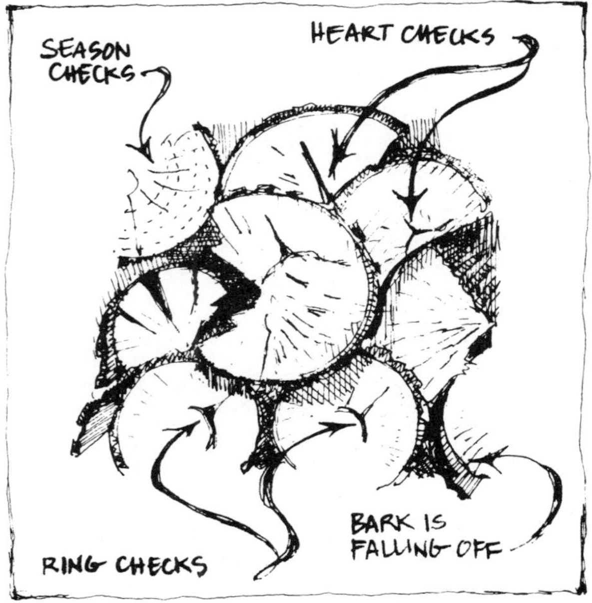

Seasoned Wood Shows Splits

FUEL VALUES OF SOME WOODS

HIGH	MEDIUM	LOW
20-27 million BTU's per cord	17.5-20 million BTU's per cord	12.6-17.5 million BTU's per cord
American beech	Black cherry	Balsam poplar
Apple	Black gum	Basswood
Black birch	Black walnut	Black willow
Black locust	Elm	Box elder
Blue beech (American hornbeam)	Gray birch	Butternut (white walnut)
Crabapple	Holly	Catalpa
Dogwood	Honey locust	Chestnut
Eucalyptus	Magnolia	Cottonwood
Hickory	Oregon ash	Largetooth aspen
Hop hornbeam (hardhack, ironwood)	Red gum	Quaking aspen (popple)
Live oak	Red maple (soft maple)	Red alder
Persimmon	Sassafras	Tulip poplar
Shadbush	Silver maple	
Shadbush	Sycamore (Buttonball, buttonwood)	
Sugar maple (hard maple)	White birch	
White ash		
White oak		
Yellow birch		

species. The reason that heat potential varies among kinds of wood is that the density of wood varies. Thus, a pound of pine and a pound of oak will each produce about 7,000 BTU's, but the pound of pine will take up more room. Since wood is sold by volume, and your stove's firebox holds a given volume, denser wood makes better firewood. It has more heat potential for a given volume.

As a general rule, hardwoods are denser than softwoods, but there are some exceptions. On the average, a cord of air-dried softwood weighs about one ton, while a cord of air-dried hardwood weighs about one-and-three-quarter tons.

There is nothing wrong with burning the lighter woods if they are what is available. In the East, many look upon softwoods with disdain because we have so much of the longer-burning hardwoods. But many of our Western friends heat their homes with woods that are low on the BTU chart. Douglas fir is common in Oregon, aspen and pine in Colorado.

There are even some advantages to burning such woods. Soft deciduous types such as poplar, basswood, or cottonwood are easy to cut and split. Their light weight makes them easy to handle and transport without expensive and complicated equipment, and they can be sawed and split by hand without great effort. These latter attributes are particularly valuable after you have labored through several cords of rock maple. Every woodburner should finish the annual fuel-processing ritual by cutting and splitting a cord of poplar or pine for the psychological benefit alone.

Be aware of the shortcomings, however. In an attempt to prolong the length of the burn when using these softer woods, many people rely on restricting the air entering the stove as much as possible. This practice is asking for trouble regardless of what wood type you are burning. A fire without an adequate air supply smolders and produces large amounts of smoke, and this is the beginning of an ideal creosote situation. Conifers such as pine or spruce contain higher resin levels than do the deciduous hardwoods, and this too will contribute to their susceptibility to result in creosote when burned.

The key to burning less-than-desirable wood is the flexibility with which you operate your stove. Burn these woods during weekends or days when you will be at home to tend the fire regularly. If you must burn them overnight, plan on getting up to refuel the stove during the night rather than reducing the air supply to the point where creosote will be a problem. Split some of these chunks much finer than others and use them when you have the doors of your stove open to enjoy the fire. They will burn cleanly and brightly. Also use thinly-split pieces as kindling to start or revive a fire that has died down. These woods will also serve nicely in the milder weather of spring and fall when a morning or evening fire is all that is needed to take the chill off.

There is a great deal of potential heat in every cord of low density wood you may have access to. Harvesting the unwanted wood as fuel can contribute greatly to your year-round heating plan when you use common sense and flexibility in your stove-tending routine.

Woods vary in other ways besides density; ease of splitting, speed of drying, resin content, ash content, aroma, tendency to throw sparks,

FUEL VALUES OF WESTERN TREES

HIGH	MEDIUM	LOW
Ash, velvet	Ash, Oregon	Alder
Dogwood	Pacific mountain	Aspen
Ironwood, desert	Birch	Cedar
Larch, western	Buckeye	Cottonwood
Locust, New Mexico	Buckthorn	Fir, Alpine
Madrona	Catalpa	balsam
Mahogany	Cherry	grand
Maple, western sugar	Chinquapin	noble
Mesquite	Cypress	white
Mulberry	Elder	Hemlock, mountain
Oak, all kinds	Fir, Douglas	Maple, Oregon bigleaf
Osage orange	Hackberry	Pine, all kinds
Soapberry	Hemlock, western	Poplar, balsam
	Juniper, Utah	Redwood
	Maple, dwarf	Spruce
	vine	Sycamore (buttonwood, plane)
	Plum	Willow
	Retama	
	Smoketree	

The Crosscut Saw

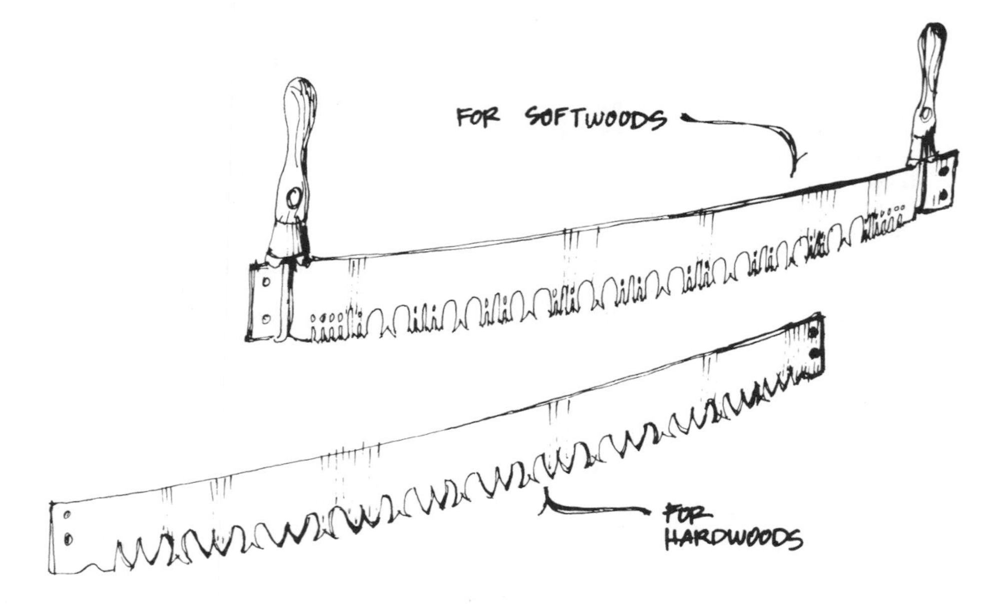

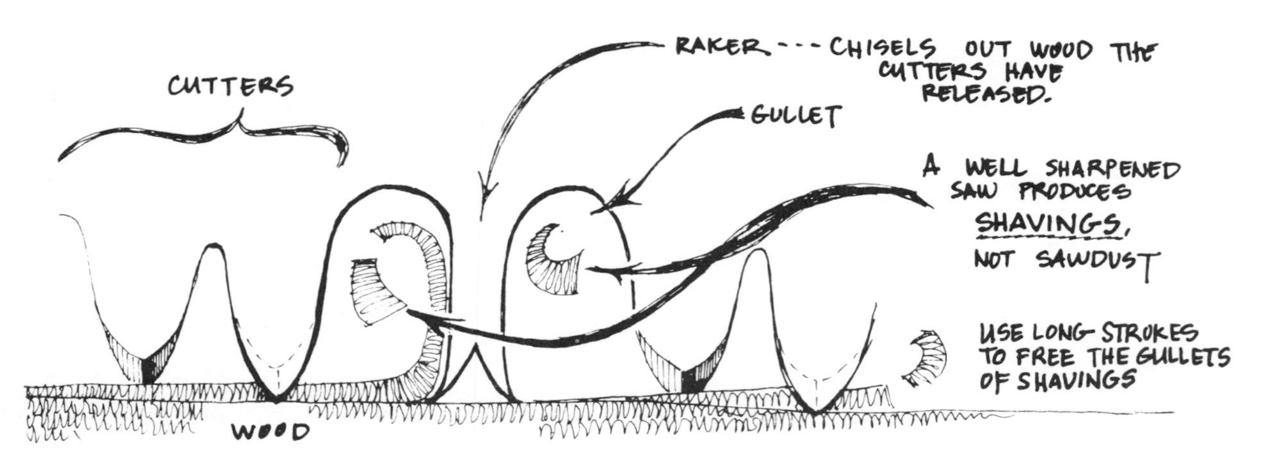

and coaling qualities. These are the factors that give each kind of wood its characteristics as firewood.

Oak can take two years to dry while ash can be burned soon after it is cut. Beech dries quickly once it is split. Birch should always be split; otherwise it can rot inside its bark instead of drying.

Apple wood has a nice aroma. If you are lucky enough to get some, save it for when you are using your stove as a fireplace. Juniper smells like gin when it burns; it has a low heat value so don't bother using it unless you like the smell of martinis.

Aspen (popple), cedar, and hemlock throw sparks. Burn them with stove doors closed, or use a firescreen.

Tamarack (larch) has the highest heat value of the softwoods, better than such hardwoods as white birch and elm. Mango, madrona, and loblolly pine are some of the woods that are unfamiliar to us here in New England but are burned in other parts of the country.

Chain Saws

The chain saw replaced the crosscut saw in the hands of professional loggers for simple reasons of productivity. Most firewood cutters today make a similar choice for the chain saw, which is usually the most expensive and substantial tool they employ in the yearly task of replenishing their fuel supply. A professional woodcutter, who depends on the saw for his economic survival, expects a well-maintained saw to cut about one million board-feet of hardwood timber before a major motor rebuilding is necessary. The more numerous, shorter cuts required to work up firewood result in a life expectancy for a good-quality chain saw of about 2,000 cords. Plan on purchasing just one; it will last you the rest of your life. If you already own a fairly new saw, don't buy another; the maintenance and operating techniques required to keep the tool in safe operating condition will also make it last longer.

The majority of professional loggers on both coasts prefer saws of Swedish or German extraction. The American-made Homelite also enjoys a good reputation among these quality-conscious consumers. Look for a saw manufacturer who offers a complete line of professional use sizes (60–120cc motors), and consider a saw in the 60–80cc range. These same manufacturers also offer saws in 30–60cc "homeowner" sizes, but a saw with a slightly larger motor will have to work less hard over its years of use and generally will be more trouble-free. The additional weight of a larger saw (most weigh 15–18 pounds) is compensated for by the smoother control it affords the operator. The motor will "lug-down" less while making a long, heavy cut, and this will require less "horsing" of the saw. The wider torque range of a large saw allows the sawyer better throttle "feel" for various chain speeds and in the long run will produce less fatigue. The poise and concentration so essential to safe operation are more easily sustained if the saw cuts "like a warm knife through butter."

Several features have emerged in the last few years to improve the quality of chain saws. *Anti-vibration handles* are strongly recommended to reduce finger-muscle fatigue. Make sure that both ends of the front and rear handles are mounted to the motor case by means of rubber bushings. This feature is worth many times its additional cost. *Electronic ignition* prolongs spark life, prevents plug fouling, and promotes evenness of idling speed. A *chain brake* effectively

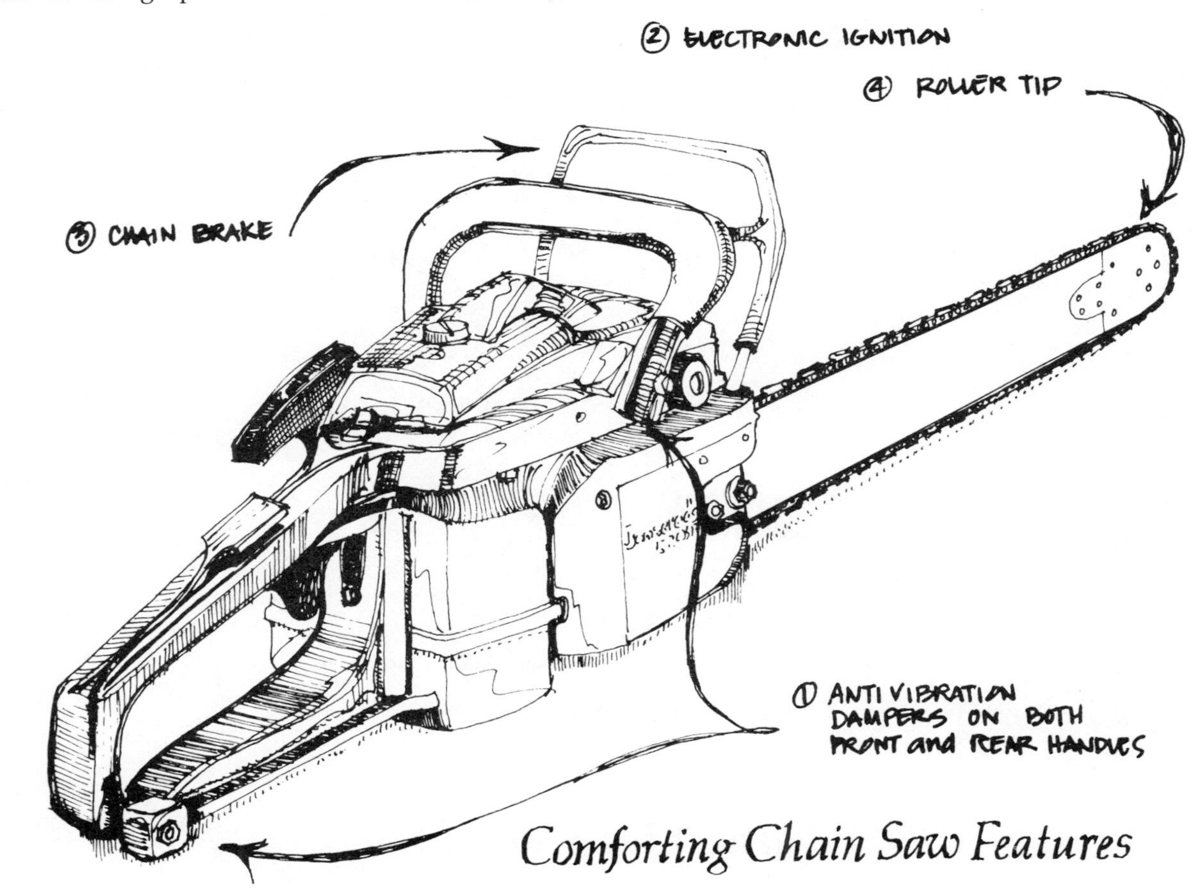

Comforting Chain Saw Features

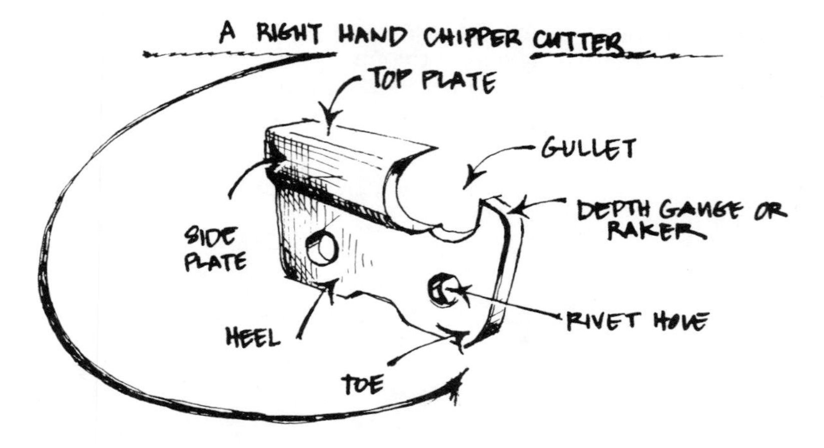

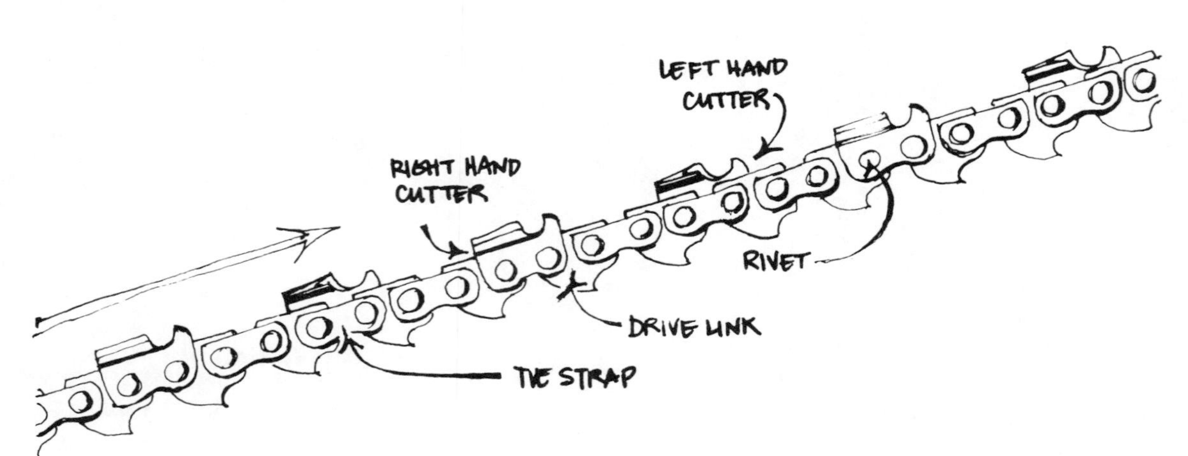

Parts of the Saw Chain

halts the chain automatically should the saw kick back. This provides safety insurance for the novice sawyer, especially when mastering plunge cuts. Needless to say, you should never allow a chain-brake to become a "crutch" that prevents you from gaining a deep respect for the danger of a kick back. Honor the correct procedures and the chain brake will never be used. *Roller-tip* guide bars, while reducing chain friction and increasing chain speed, can be unreliable, especially if used heavily for felling cuts where a severe pinch can ruin them. Twigs occasionally catch in the small gap where the wheel emerges from the fixed portion of the bar, throwing the chain. If used mostly for bucking, or sawing-off cuts, roller tips are an asset.

Saw Maintenance

The chain saw has been described as the most dangerous tool used by non-professionals. The importance of proper chain sharpness for safe operation cannot be overemphasized. Experienced wood cutters monitor the wood chips produced by each cut as a guide to chain sharpness. The larger the individual size of the chip, the sharper the chain. A dull chain produces sawdust. It helps to know something about how the chain cuts in order to maintain it well.

The teeth of the chipper-chain are designed to strike a balance between two somewhat contradictory actions; a gouging or "pecking," and a slicing or shaving of wood fibers. As the forward part of the tooth link, called the depth gauge, or raker, strikes wood fiber, it levers the rear portion of the link with its cutting edge away from its track in the guide bar, and drives it into the wood in an arcking path. If this arc is too shallow, the cutting edge merely nips the fibers, if too deep, it claws roughly; if correct, a smooth slice is drawn. Needless to say, if the cutting edge itself isn't sufficiently sharp, not much of anything happens. The height of the raker determines the depth of the cutting arc, and should be filed to six-tenths of the height of the cutting edge. A gauge plate that fits over the tooth is used to check this. The cutting edge on a new chain, factory-sharpened with high speed automated equipment, often provides only minimally acceptable sharpness and should be "dressed" with a few light file strokes. No light should reflect from the edge of a truly sharp tooth. One soft lick with a flat file on each raker and a new chain is ready to chip.

Remove lubricating oil from the teeth before filing. The filings from dry teeth will not cling to the file as much, and this makes the process more effective. Drying the teeth is accomplished by beginning a cut into a piece of wood, then abruptly shutting-off the saw's motor with the chain remaining in the cut.

When sharpening the chain, the file should be rotated slightly after every two or three strokes and wiped or brushed frequently to provide "clean" abrasion to the steel of the tooth. While the actual file stroke is best rendered in whatever manner works for the individual, experienced sharpeners will hold the file in one hand and force the forefinger of the other to steady the tooth by pressing on its backside, toward the file-stroke and down into the guide-bar groove. Various types of file-holders and guides can be useful to the novice sharpener in establishing "feel" and tooth-to-tooth consistency, although these devices are typically abandoned with gained experience, since they tend to obstruct visual scrutiny of the emerging cutting edge.

A little guide-bar maintenance goes a long way to reduce wear on the bottom of the chain. Flip the bar over after every ten hours of use to promote evenness of wear, and file off any ridges that may have formed on the outer edge of the rails. Take care to preserve the curvature of the Stellite-hardened tip by using sweeping strokes and light pressure. The heft of a 10-inch mill bastard flat file aids this task. The 6-inch flat file is

Guide Bar Maintenance

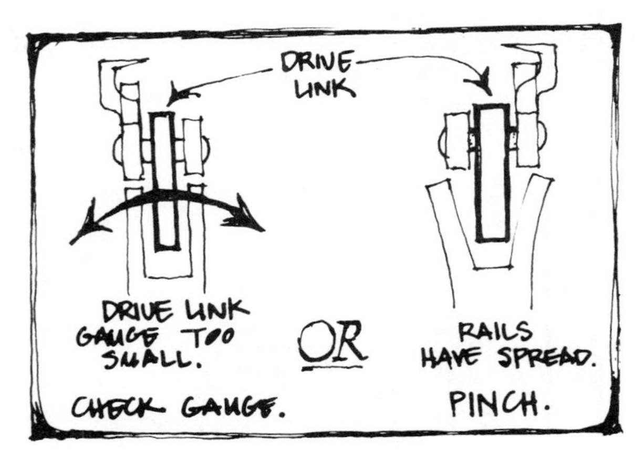

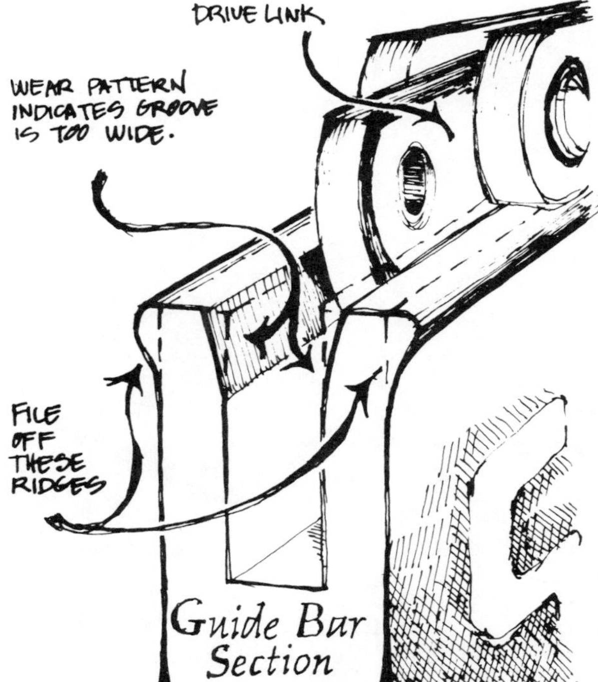

Guide Bar Section

best saved for the tooth rakers. Rapid continuous wear of the inner rail edges indicates that the groove is too wide. If localized, this wear results from over-spreading rail pinches; the groove should be narrowed with a hammer or vise. (To avoid over-spreading, introduce a thick screw-driver blade to the pinch at an angle that allows the taper of the blade to barely clear an unpinched portion of the groove, then drive the blade along the groove through the pinched area with even, successive hammer-taps.)

If rapid wear is constant around the entire bar, the chain's drive link thickness is too narrow for the groove. Standard 3/8" chain is available in drive link widths of .043, .048, and .052 inches. When replacing a chain, be sure to obtain the proper match for the bar.

When purchasing a new bar, consider having your old one reconditioned to be available as a spare. This costs about half the price of a new bar and wears for about half as long. Many saw repair shops that trade with professional loggers offer this service. Keep in mind that shorter bars afford greater chain rpm's, yielding more power and efficiency. Saws with 60 to 80 cc motors are most comfortable with 16 to 20 inch bars.

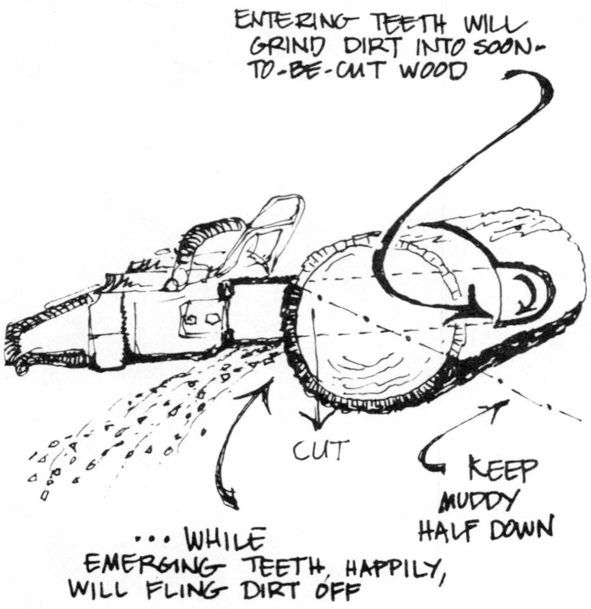

Keep Entering Teeth in Clean Wood

Attention to bar lubrication will also prolong chain life. Some shops will draw bulk "cling" type bar oil into your own container, and this is an inexpensive way to buy it. True cling oil will pull from one's fingertips in long, taffy-like strands. Used automobile or tractor crankcase oil, decanted to remove sludge, has not been known to harm the oil pumps of today's high-quality saws over the entire life of their motors and can represent considerable savings, especially when a saw is faced with a five-cord truckload of muddy logs that will consume at least one chain loop anyway. For those who wince at this suggestion, another alternative is to purchase inexpensive 40-weight motor oil by the two-gallon jug.

Although the cleanest piece of firewood contains in its fibers trace amounts of minerals that will eventually dull the chain that cuts it, mud on bark is the biggest enemy of chain sharpness. When bucking a log that is dirty halfway around its circumference, roll the log so that the dirt is on its bottom half. Thus, when cutting through it from the top, the chain teeth emerging from the bottom half fling the dirt off the log, rather than grinding it into yet-to-be-cut wood fibers. Should a log be very dirty, trim the bark with a hatchet or small axe from one quarter of the circumference along the intended cutting line, introducing a clean surface for a plunge-cut to be made directly through the log. The saw can be first directed to cut upward through the top half of the log, then downward through the bottom half. The chain teeth will then always engage the dirty bark from the inside out, throwing the dirt into the air. These attentions pay off in filing time and chain life.

Splitting Logs

The process of striking a piece of wood sufficiently hard with an instrument so that it will fall into two pieces has stimulated countless hours of discussion on technique and tool. One can even read an entire book devoted to the subject. The truth is, though, that there really isn't a great deal you need to know about splitting wood. It is sufficient to know the reasons why you must split wood at all, and to be familiar with the splitting tool options available to you.

There are two reasons that large chunks of wood must be split: The most obvious is to reduce the size of the chunk so that it will fit into your particular stove. The second is to facilitate drying. The more surface area of the wood that is exposed to the sun or air, the faster it will dry. Some types are especially difficult to dry unless split, particularly members of the birch family. Indians favored birch bark canoes because water would not easily permeate the tough white layers. The same characteristic keeps the wood from drying well unless split, even the smaller pieces.

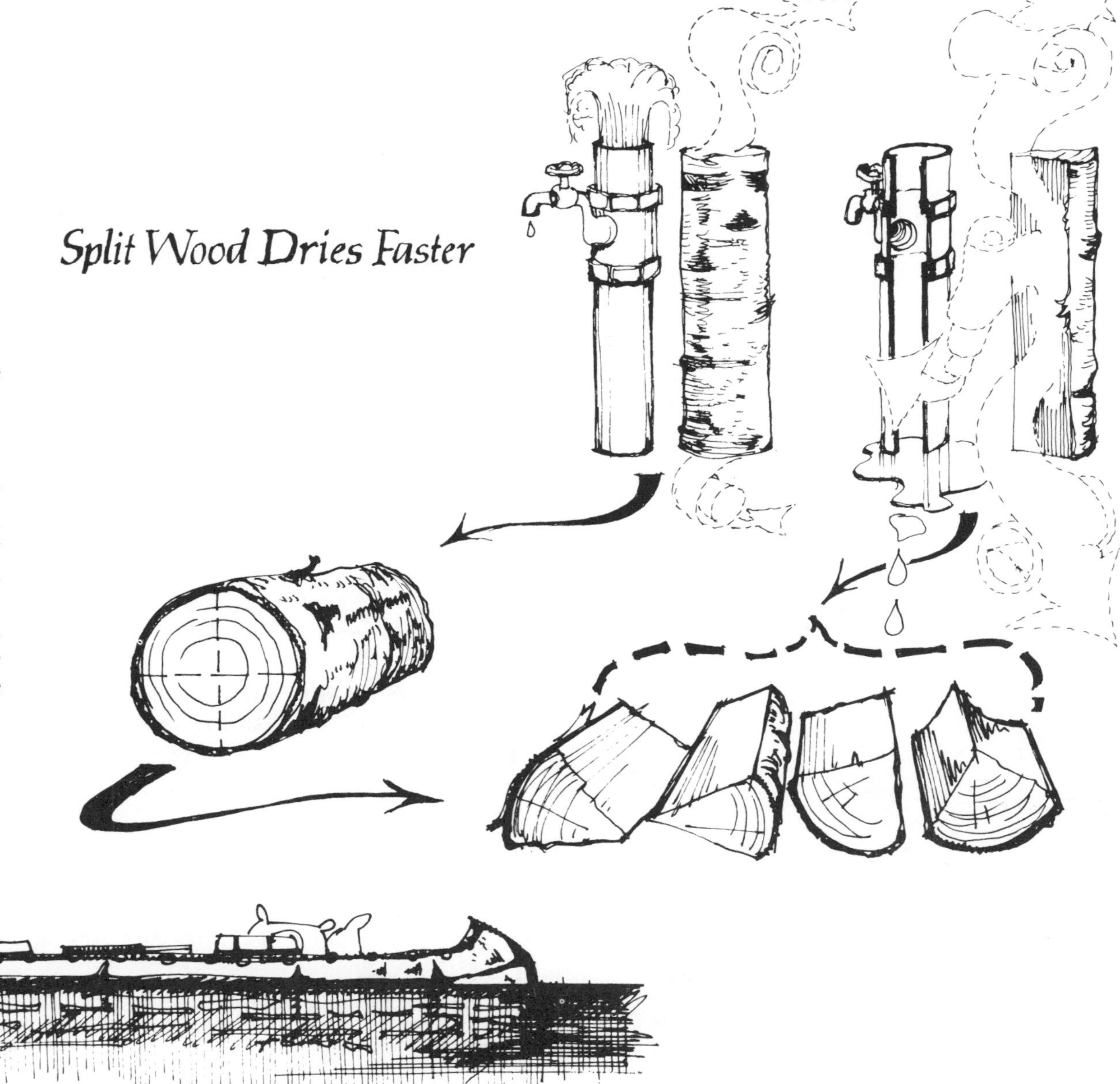

Maintain Your Taper

AXE HEAD

SOFTWOOD GAUGE

HARDWOOD GAUGE. TAPER IS LESS ACUTE, WITHSTANDS HARDER ABUSE.

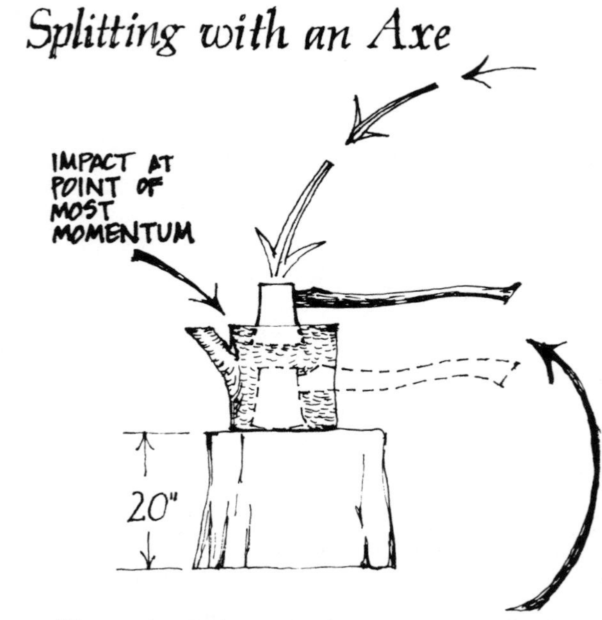

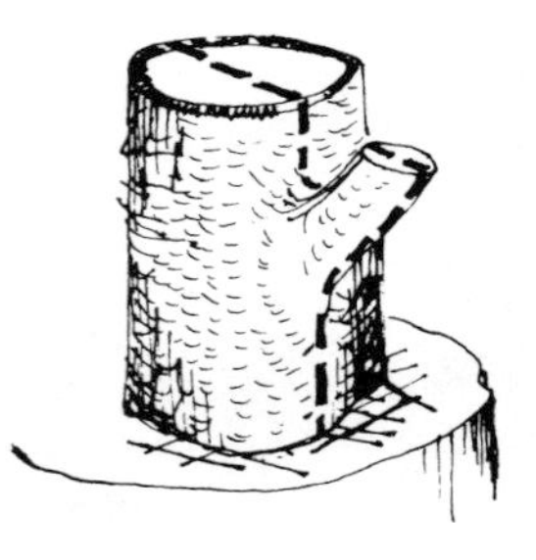

MAKE THE FIRST SPLIT THROUGH THE BRANCH STUB.

You will need to decide whether to split the wood by hand or to rent, borrow, or buy a power splitter; there are advantages to each. Splitting wood by hand is quiet, clean, and good exercise. With certain sizes of wood and adequate ambition on the part of the splitter, hand splitting can get the job done more quickly than the hydraulic approach. Splitting wood by hand with one or more partners can make the work go even more quickly as well as to make it easier and more social. A team may work together with two tools, or one may specialize in setting the pieces on end while the other does the actual splitting. This latter method is a good way for husband and wife or parent and child to work together.

A power splitter is noisy, dirty, and more expensive than a hand splitting procedure. Don't expect to speak in anything but a holler at your companions, and don't plan to enjoy the sounds of spring birds. You will smell of gasoline and exhaust at the day's end. There is a greater risk of accidental injury when you split wood with a powered splitter, and the process of bending and lifting each piece a foot or two into the air can put a strain on your back. On the other hand, an hydraulic splitter is relentless in the amount of work it can accomplish in a weekend. It does not need to stop periodically to be fed beer, nor will it pause frequently to admire the work it has done so far. It is a tireless laborer.

As with many decisions you make in life, the choice of how to split your wood may best be answered as a compromise. Split everything you can by hand for the sheer pleasure of the activity. the knotty chunks and pieces of elm that remain will make short work for a powerful hydraulic splitter.

Hand Tools

You will find any number of hand splitting tools available, all claiming to have minimized the effort required to make two pieces of wood out of one. Some even make four pieces with one stroke. There are many materials and designs to choose from, and it is very likely that you will find each of your neighbors favoring a different device. It will be worth your while to try each neighbor's preference over a period of time and then draw your own conclusions. To begin with, start your collection with a good splitting maul. Select one with a high grade, heat treated steel head that is wide enough to keep from getting jammed in the wood. The heads come in different weight sizes, but a seven-pound head is about right. Look for one with a smooth, straight grained hardwood handle for durability. Southern hickory is popular for handles. The handle length should be as close to thirty-two inches as possible.

A kindling axe will be needed for smaller work. Buy a kindling axe with a one-and-three-quarter-pound head and eighteen-inch handle.

A third tool for your beginning collection is the three-and-a-half-pound axe. With a twenty-eight-inch hickory handle, this is the right size for those pieces of wood that fall between the sizes best suited for the maul or kindling axe.

The Power Splitter

An hydraulic splitting machine may be purchased, rented, or borrowed. Since these machines are expensive, the latter two choices are the most feasible.

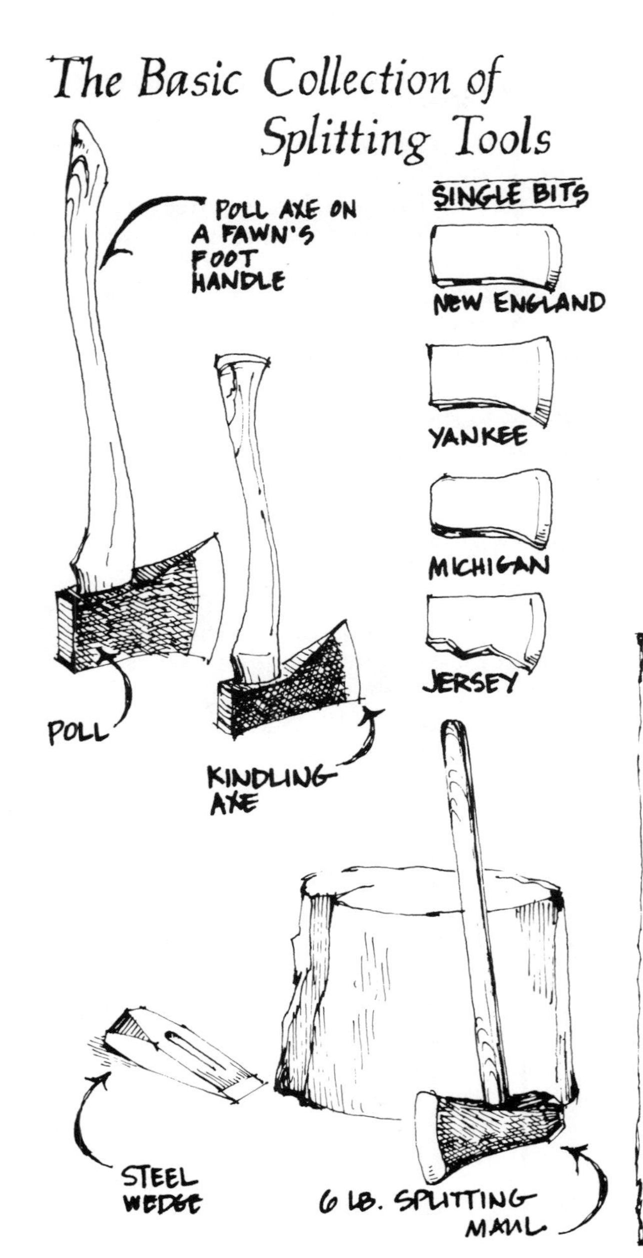

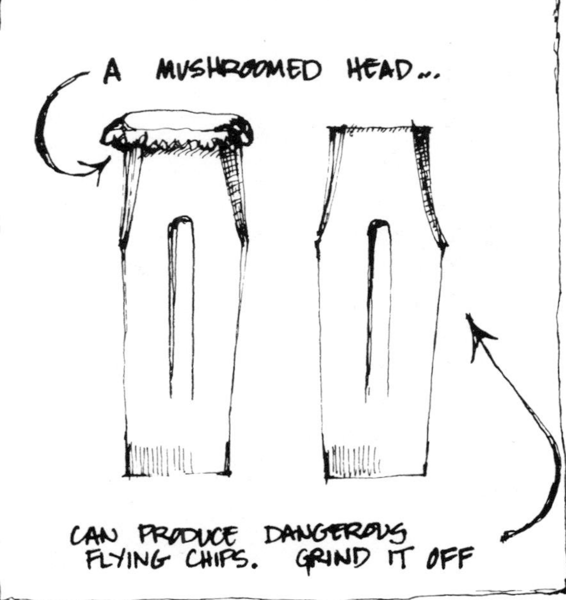

Plan to reserve the machine in advance of the days you will want to use it. You might find an arrangement where delivery of the splitter to your home is included in the rental fee. Otherwise you will need to pick it up. This requires a trailer hitch on your car. Be sure to confirm in advance that the size of the ball coupler on your hitch is the size required by the splitter.

Don't let the owner of the splitter leave you alone with it without showing you how it operates. Have him check the oil, gas, and hydraulic fluid, and then ask him to show you how to start it. You will want to have oil, gas, and fluid on hand as well. Check the level of the oil and hydraulic fluid each time you fill the gas tank. You will also need grease for the beam the ram rides on.

To split wood efficiently, there should be more than one doing the job. The splitter should be positioned as close to the wood as possible, with enough room to maneuver. A good system can be developed with four people: one to operate the lever, another to load wood on the splitter, the third to make sure there is wood in position to load and the fourth to throw the split pieces out. Sometimes the wood is heavy enough so that two people must load and hold. Also, with the four-person system, the positions can be rotated every half hour to prevent tiring.

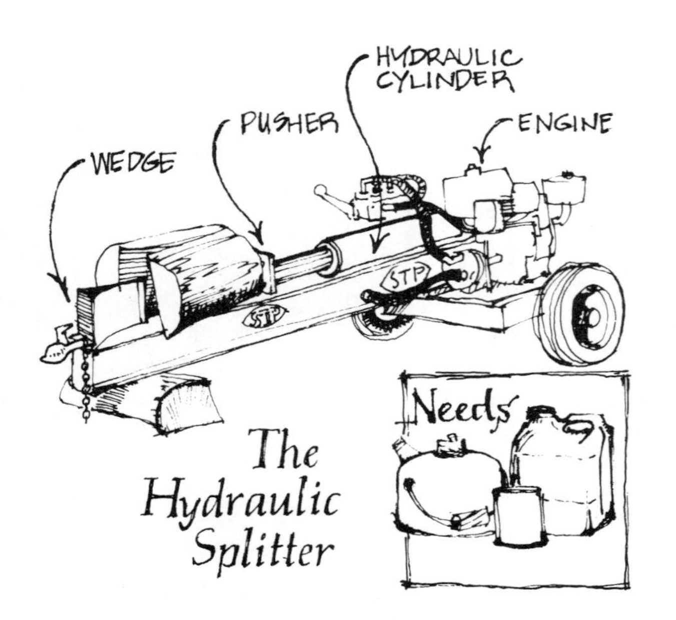

The whole idea of the splitter is to save time and effort, and initial wood position is important. Get the wood as close to its final destination as possible before splitting. This will save travel time to stacking location.

Whenever operating any machinery, safety procedures must be observed. Very seldom will the machine cause an accident; it is the people operating it. Before starting the splitting operation, agree on safety procedures. Signals should be determined between operator and loader. Loaders must keep hands out of the way of the ram at all times. The operator of the lever must watch the loader's hands carefully; this job is the key in preventing accidents. Always keep an eye on wood being split; sometimes dry wood will pop and kick out. Let the machine do the work; if some chunks do not split completely, back off the ram, turn the wood over, and run it through again.

Occasionally a large chunk of wood will become jammed on the wedge. The first inclination is to get out a sledge hammer and begin banging away, but this is not letting the machine do the work. Instead, secure one end of a chain around the impaled wood and the other end around the ram. Put the splitter in reverse and let the ram pull the wood away from the wedge.

Certain safety items should be employed when using a splitter. Steel-toed shoes are a top priority. It never fails but a log will drop on someone's foot, always provoking anger and sometimes causing injury. Snug-fitting gloves are essential. Safety glasses should be worn as well as ear protectors.

The Drying Process

One of the first lessons a woodburner learns is that firewood needs to be seasoned in order to burn well. If you have ever run out of seasoned firewood and tried to burn freshly cut logs, you know that lighting green wood is like trying to kindle a zucchini squash. Once you do get it started, it usually burns with a cool, smoky fire that is apt to cause a great deal of creosote in your chimney. In addition, much of the heat value of the wood is wasted in vaporizing the moisture it contains.

"Green" is a term often used to describe an inexperienced ballplayer or a multitude of other apprentices just launching a career. All have possibilities, but they simply haven't reached their full potential. Green wood is like that. It has a lot of potential, but it isn't ready to burn.

The greenness or dryness of wood is usually expressed in terms of its moisture content. Moisture content is the ratio of water to dry wood by weight. Thus, a ten-pound log with a moisture content of 100% is not made totally of water, but contains five pounds of wood and five pounds of water.

A freshly cut tree can have a moisture content of 50% to 150%, depending on the species and the time of year it is cut. Ash and beech, for example, contain less than half as much water as oak when green. Trees cut in winter or early spring before the sap begins to rise have a lower moisture content than those with the sap flowing. On the other hand, trees with leaves will lose some of their moisture content if left whole on the ground after they have been cut, because the leaves draw water from the wood.

To give an example of the heat you lose by burning green wood, let's use the ten-pound log containing five pounds of wood and five pounds of water. Each pound of water is roughly equiva-

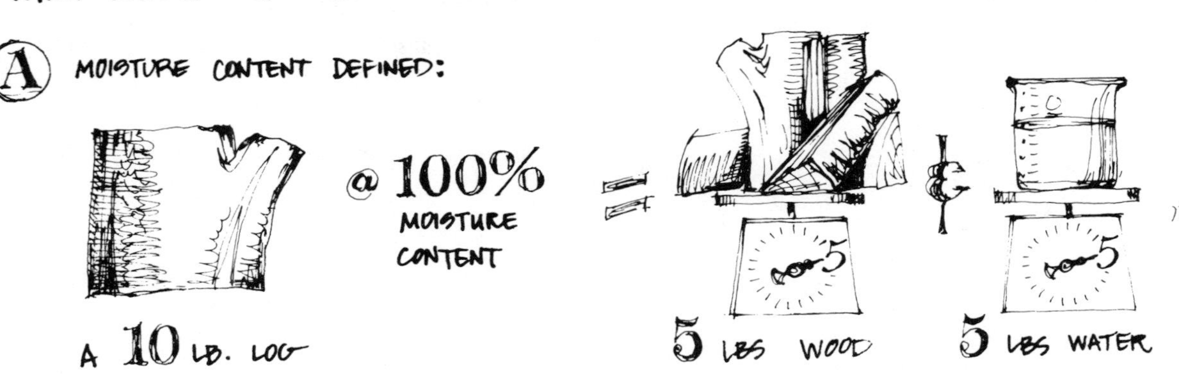

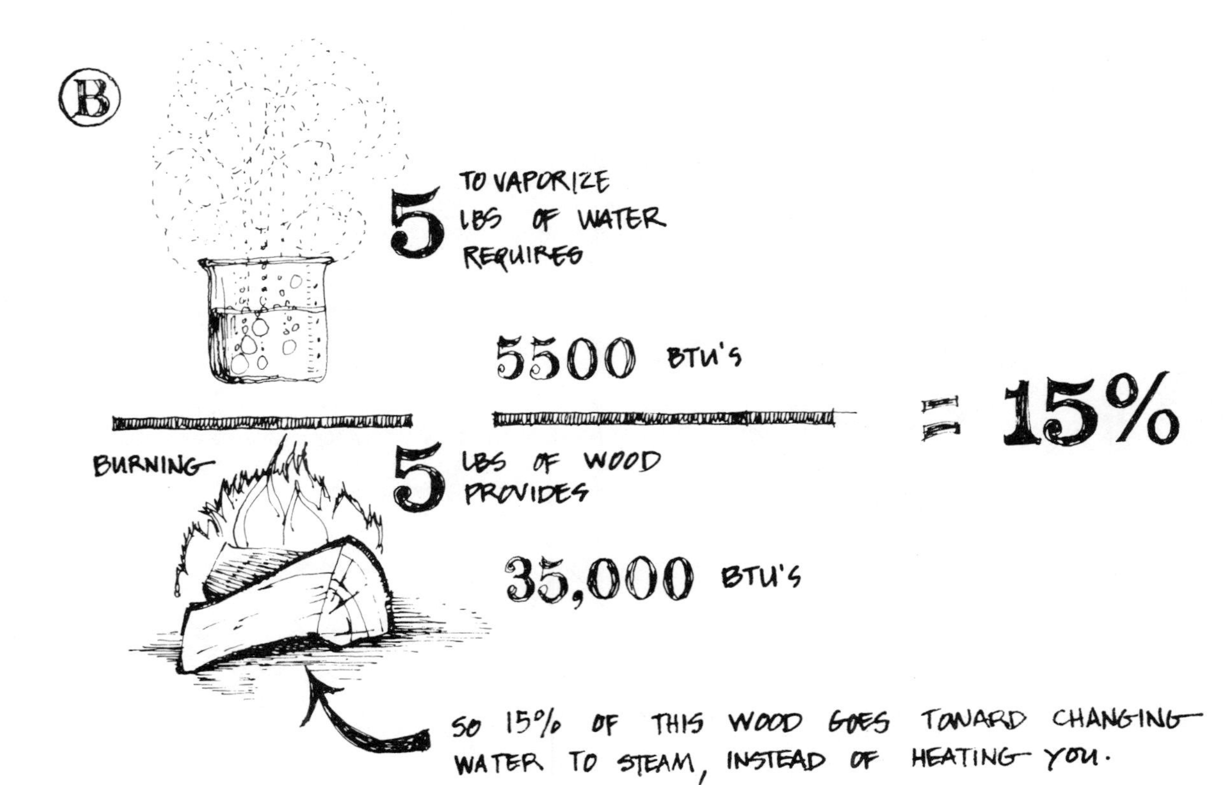

Learn From the "Experts"

If we could do something this stupid, we figure you could as well. Save yourself the trouble. But don't let the humorous tone deceive you; learn from our mistakes.

A Tale of Two Ashes
Part 1

"*I was living in a rented house with another new Vermont Castings' employee while we were waiting to move our families to Vermont. Two men living alone are not always the best of housekeepers, so when your wife is visiting for the weekend, you make a frantic last-ditch attempt to neaten up. In our haste we vacuumed everything with an ancient Electrolux —floors, rugs, kitchen table, and even the woodstove. The weather had been mild, so the stove had been cold for several days. The telephone rang. I turned off the vacuum and went to answer it. As I was talking I smelled something that reminded me of smoke. Because a neighbor had been burning leaves that afternoon I thought nothing of it and kept talking. By the time I hung up the smell was much stronger. I went back to finish vacuuming. On the hearth was the trusty Electrolux, belching the most acrid, foul-smelling smoke I have ever encountered. Some fast action averted disaster, but my credentials as both a novice stove operator and poor housekeeper were firmly established. Worse still, when I had to explain the charred Electrolux to my landlady, I felt like a complete "ash."*"

Dwight Stimson

lent to one pint of water and it takes about 1100 BTU's of energy to vaporize each pint. That is 5500 BTU's are needed to drive the five pounds of water from the log. The heat value of the five pounds of wood is about 35,000 BTU's. Thus about 15% of the heat value is lost.

You should try to have your wood split and stacked at least six months before you will need to use it. A year ahead is even better, but sometimes this is not possible. If you find yourself getting too close to the heating season with a pile of green wood, there are steps you can take to get your firewood reasonably dry. First of all, it helps to be aware of the theory behind drying wood, so that you can apply it to speed up the process. Even if you get your wood in early, it is useful to understand how wood dries so that you can set up your woodpile for best drying efficiency.

How Wood Dries

Wood dries when air passing over its surface absorbs moisture and carries it away. The reason this occurs is that moisture will always travel from an area of high concentration to an area of low concentration until the moisture level is equal between the two areas. Evaporation of moisture from the wood surface creates a capillary action within the log that draws water from the wood cells. By this process the "free" water located inside the wood cells is drawn out, followed by the moisture that is trapped in the walls of the cells. Wood cells have passageways between their ends, and for this reason moisture travels faster to the end of the log than out through the bark.

When the free water has evaporated, the wood is at approximately 30% moisture content. At this point, cell wall shrinkage begins to occur and you will start to see check splits at the ends of your logs. The rate at which the drying process occurs will depend on several factors.

The first is the relative humidity of the air around the wood. Relative humidity is a measure of the moisture in the air at a given temperature. It indicates how much more moisture the air can absorb before the moisture condenses. The lower the relative humidity around your wood pile, the greater the evaporation of moisture into the air.

The higher the temperature of the air, the more water vapor it can hold. Increasing the temperature without increasing the moisture content of the air around the wood pile will decrease the relative humidity of the air and speed up the drying process.

When radiant energy from the sun, a radiator, or a wood stove strikes an object such as a log, the surface temperature of the object rises. The higher the surface temperature, the faster the evaporation from this surface, as long as the relative humidity of the air is lower than the wood's moisture content.

Air circulation is important, too, because air absorbs and carries away moisture from the damp wood surfaces: The more air movement or convection currents there are throughout the wood pile, the faster the drying will be.

Size, shape, and density of the wood pieces all play a part as well. The greater the exposed

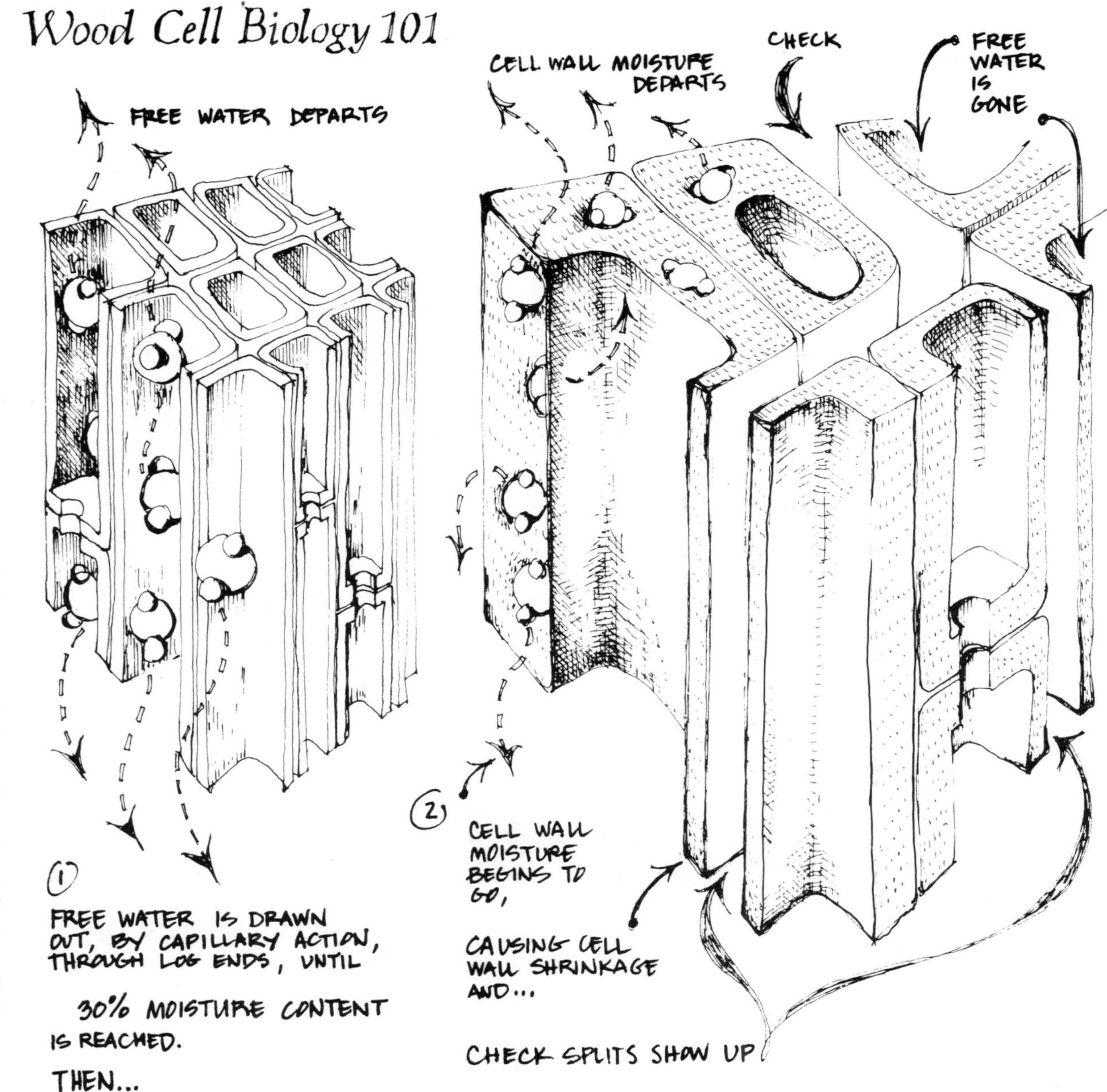

The Well-dressed Logger

The well-dressed logger dresses for safety, mobility, and comfort in that order. The extremities must be protected. This means that high work boots, snug gloves, and a hard hat are requirements. Safety glasses and ear protectors are worn by professionals and are equally suitable for the weekend do-it-yourselfer. Ballistic chaps, which will protect your legs, are helpful, but cumbersome. The need for them can be obviated by attention to safe operating procedures.

Other clothes should be warm and loose, but never baggy. Loose cuffs have a tendency to find their way into moving chain teeth, dragging a much-needed portion of your body right behind. Do not underestimate the importance of comfort. A warm, unfatigued lumberjack is less likely to find himself in a hazardous situation and is more able to extricate himself if he does than a cold, tired one.

A falling tree often sends a cascade of branches through the air. Even the best dressed logger can find himself stunned by one of these missiles, appropriately known as "widowmakers." Never work alone in the woods. Your protective clothing and safe operating techniques will pale beside the importance of a good friend if you should encounter trouble.

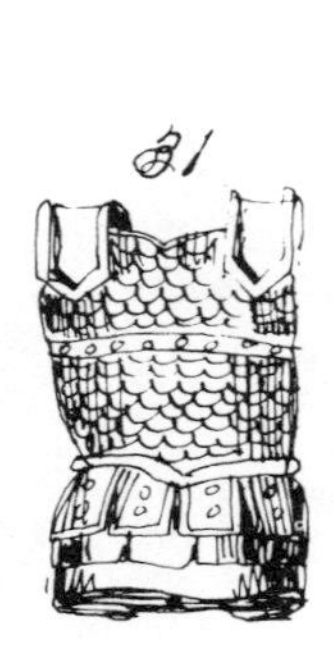

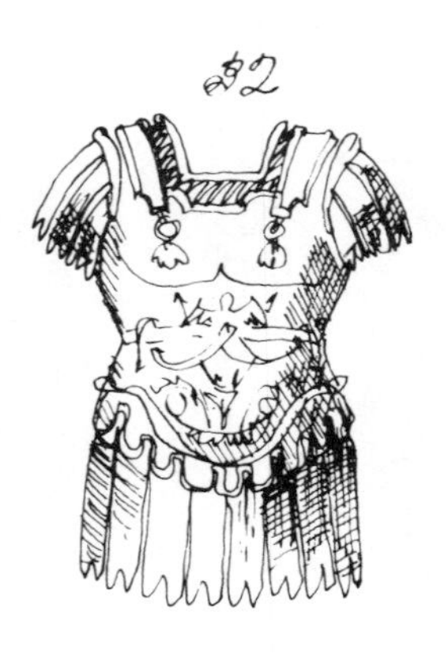

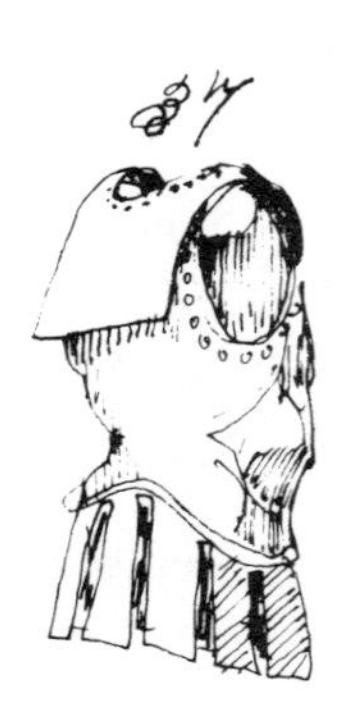

surface area of the wood, the greater the moisture transfer will be. The shorter the distance the moisture has to travel, the more rapid the drying will be. Therefore, split wood dries faster than whole logs and small pieces faster than large pieces. Denser woods will hold more moisture per unit volume than lighter, less dense wood. The greater the surface-to-volume ratio and the lower the wood density, the faster the wood is likely to dry.

Studies have shown that logs cut to one-foot lengths dried up to three times as fast as others cut to four-foot lengths. Similarly, splitting the logs cut the drying time nearly in half. Two-foot split logs took from 3½ to 6 months to reach an acceptable moisture level (under 25%), while most of the four-foot unsplit logs were still too moist after 7½ months. These logs were red maple, white birch, and red oak, 5½ to 7½ inches in diameter. Split in half, these made fairly small size logs.

To Dry Your Wood

You can use these principles of drying to shorten the time it takes to season next winter's firewood supply. Consider the following suggestions when you plan your woodpile.

Choose a location that is on dry ground and away from trees or other objects that will shade the pile and block solar radiation. Avoid stacking the wood against the side of a building, or in any place that would impede air circulation through the pile.

Build a log or concrete foundation so that the wood is raised several inches off the ground. Old wooden pallets make a good foundation if you can locate them. This will allow convection currents to flow under and through the pile. Remove any brush and high grass from under and around the foundation.

Cut your firewood pieces to stove length and split them. The closer you are getting to the heating season, the smaller you should split your wood. Remember, though, that smaller pieces will not hold a fire as long as larger ones.

Stack the wood with plenty of air spaces between the pieces. This will take up more room in your yard, but the results are worth it. Best of all is to stack in a criss-cross style that provides maximum air space. Place the pile so that prevailing winds flow into its long side.

Cover the top of the pile with plastic, a tarp, or roofing material. This will keep the rain water from leaking down through the pile and slowing down the drying process. Never wrap plastic tightly over the top and down the sides of the pile without vents, as this will block air flow through the pile.

Using these techniques and splitting your wood quite small—about three to five inches in diameter—will allow you to dry freshly cut wood in six to eight weeks, at least dry enough to light easily. If you leave the wood in bigger chunks it will take four to six months. Interior portions of the logs will still contain moisture, but the outer surfaces will be dry enough so that kindling the fire will be easy.

Wood is dry enough to burn when it reaches a moisture content of 25%. You know it is dry when end checks and cracks appear and the wood feels lighter.

Transporting Wood

Whether you buy wood from someone else or have a free or low-cost source, transportation is a major consideration. The distance may be as short as fifty or a hundred feet to move the woodpile into the basement, or a few miles from a commercial woodyard. In either case, the cost of the transportation should be kept to a minimum.

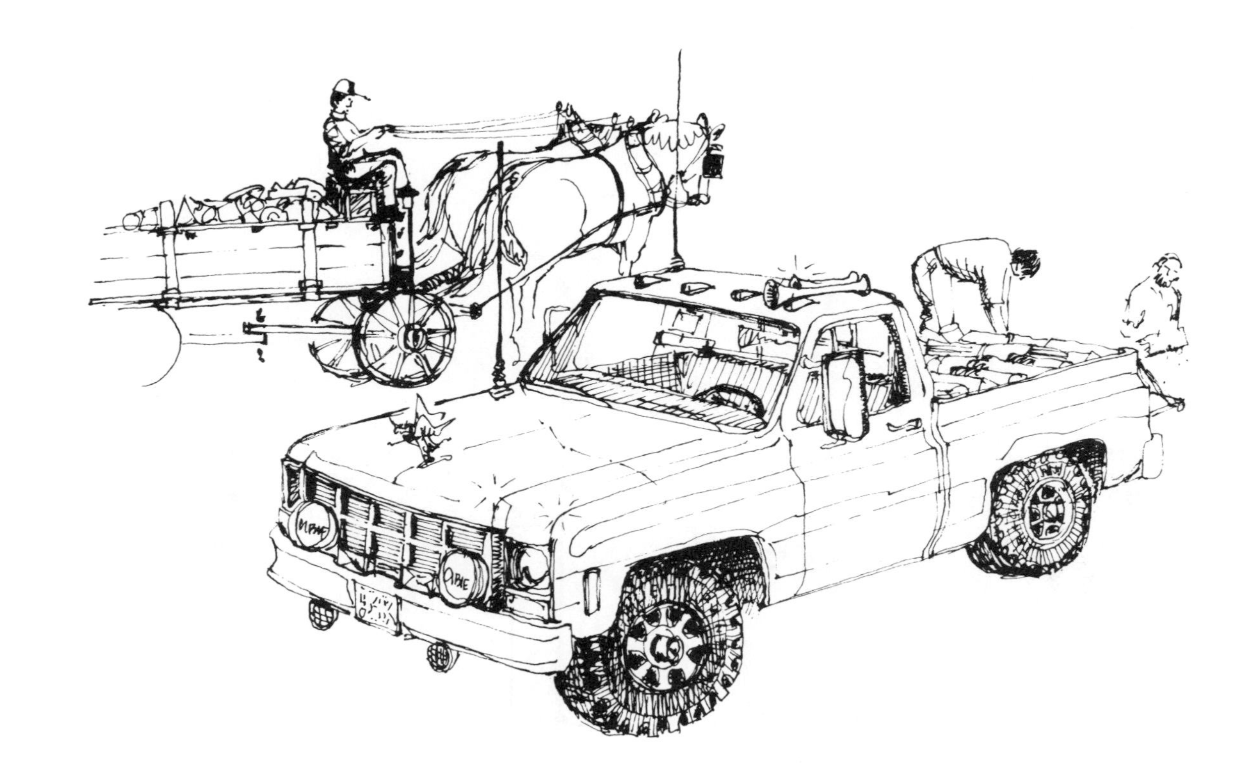

A local wood salesman learned a hard lesson about wood transportation a few years ago, even though he had the right idea. He came to our attention when one of our Customer Relations staff responded to a newspaper ad that offered cut, split, and delivered hardwood at a very attractive price. A few days later the first of the load arrived—not in the sturdy truck you might expect, but in an aged and sagging Volvo station wagon. The order was never completed, since on the next load the abused car abruptly retired from the wood hauling business after a major collapse of the suspension system. The price on the wood had been low because the seller was not making payments on an expensive truck. While the plan was ultimately self-defeating, the principle of keeping the wood price as low as possible by minimizing transportation costs was a good one. Avoiding a major special vehicle expense is one way to do that.

Many wood burners can't afford to buy the traditional pickup truck that comes to mind when you think of transporting wood. A vehicle in good condition will cost several thousand dollars, and this amount will eat up several years of the savings you are going to make on your fuel bill by burning wood. Unless you can justify the expense on other grounds it probably isn't worth it. You might find an older truck for only a few hundred dollars, but there is a good chance that the vehicle in this price range has seen hard service and will not withstand the abuse handed out in the wood-hauling process.

An ideal solution that is often overshadowed by the ruggedly romantic notion of your own pickup truck is to invest in a good utility trailer. These trailers can be hitched directly to the family car and can carry a gratifying amount of wood . . . not as much as a pickup, perhaps, but enough to make it a viable alternative.

Trailers may be purchased brand-new for around $500. Sturdy trailer frames can be had for less but require you to install the bed and sides. You may also find a used trailer for under $100. Vermont trailers are made from other old cars and trucks, and sometimes from abandoned manure spreaders.

A Utility Trailer

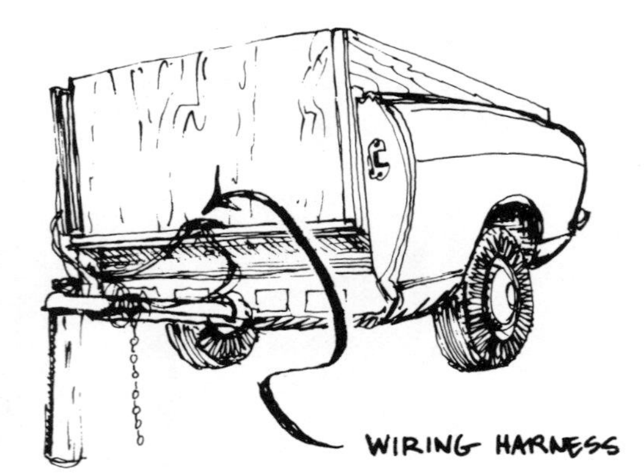

A trailer with a bed that is six feet long, four feet wide, and a couple of feet high has forty-eight cubic feet and will hold around one quarter of a cord of wood. Look for one that can carry that much weight, usually 1000 pounds or so. The owner's manual of your car will give you specific towing capacity data for your particular model. Your trailer will need to have lights and to be registered if you will be using it on public highways, but the cost for this is minimal.

The lucky woodburner for whom money is not an issue has many choices: Fourwheel drive vehicles are very popular, as are older model farm tractors. Draft horses have a special appeal, but one must be experienced in handling animals of this size. A unique machine called a Quadractor combines traction and maneuverability and is a good choice for pulling out logs. You can also pay someone else to use their machine to haul wood out of your lot.

If you are using equipment to drag logs out of the woods, work when the ground is frozen. Logs dragged on soft ground or through mud will dull your saw in minutes.

Making Life Easy

Sooner or later, every woodburner starts to develop systems. There is a system for processing wood into the proper size pieces, a certain way of transporting the wood, and a special method of keeping the stove fed. Many woodburners have found specific items to be a great help in getting the nicely stacked back yard or basement woodpile into their stoves.

Quadractor

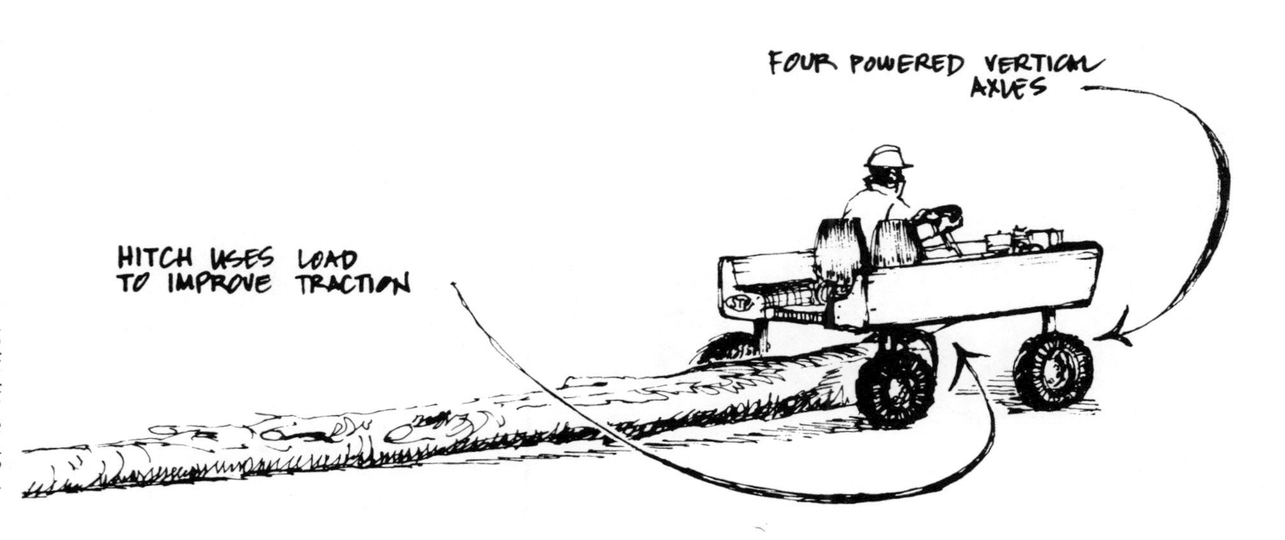

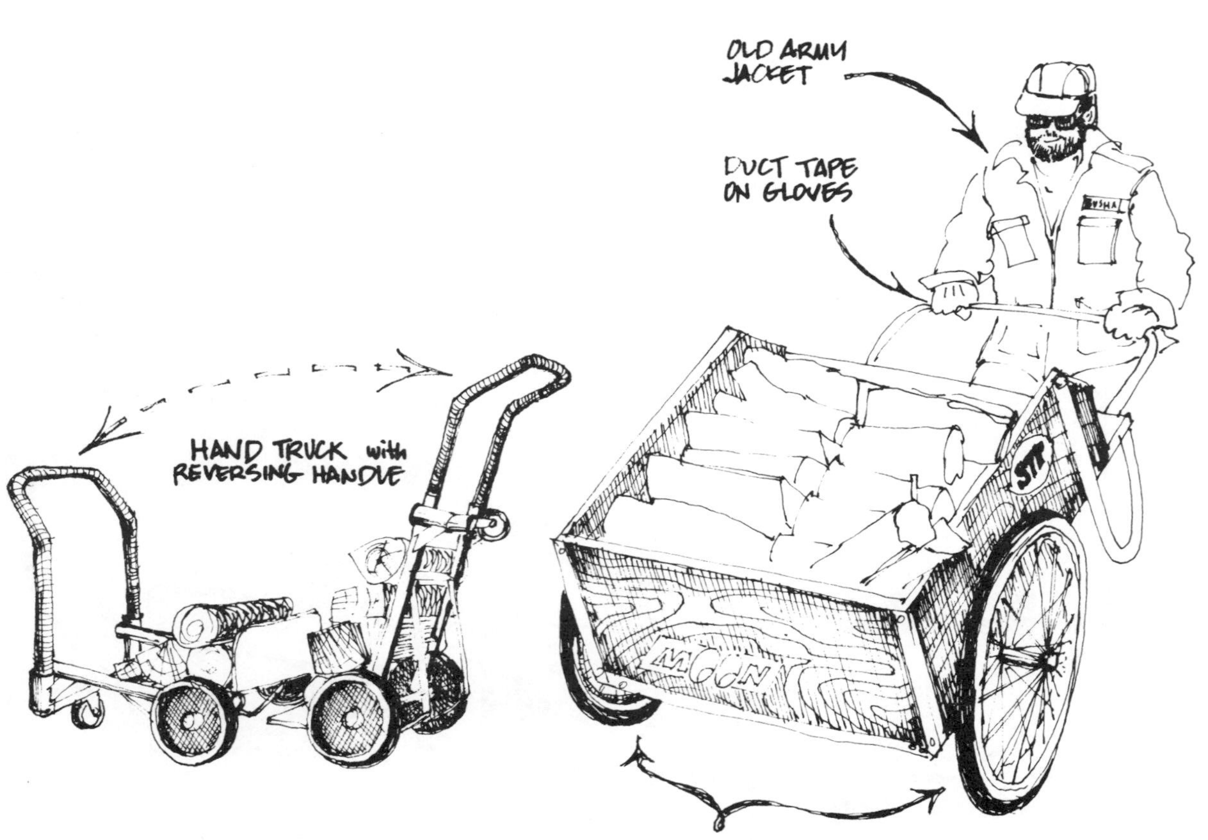

The first consideration is proper clothing. The normal inclination is to throw on whatever is handy. If what happens to be most frequently handy is your expensive down vest, you soon learn that the characteristic pose of the wood-burner features both arms pressing several chunks of wood to the section of the body between the waist and the neck. Since this is the same territory covered by the down vest, it isn't long before you realize that wood has splinters that tear nylon and that wood contains traces of dirt that leave indelible stains. An old army jacket makes excellent woodpile attire, but any garment that is sturdy and past the stage where it can be worn to impress people will do.

The right kind of gloves are also important. Avoid those that are entirely cloth; they do not grab the wood well. Instead, look for a pair with a palm material that creates more friction. Leather gloves work nicely as do cotton gloves with leather palms. A favorite style is the snug-fitting cotton glove with light plastic material for the palm. They are comfortable and provide a good grip.

A weekend of handling wood can wear out a new pair of gloves, so keep an extra pair in reserve. If you prefer a more thrifty consumer ethic, you might try what many Vermonters find to be a good technique to prolong the life of a pair of gloves; patch holes with duct tape. The tape is very sturdy and grips well.

If your woodpile is a flight of stairs below your stove, you will want a canvas log carrier. Quality log carriers made from heavy duck, nylon or leather are available just about everywhere. The carrier should have seams and hems that are reinforced with heavy stiching. Handles should be positioned so that the bag cannot tip from an off-balanced load of wood as you carry it. Two or three trips should give you a day's supply of fuel.

If your fuel source is a short distance from your house a cart will be a blessing. A wheelbarrow will do, but a two-wheeled wooden cart is ideal. It will hold an impressive amount of wood and is easily handled. While the price of the cart may equal a couple of cords of wood, you will find the investment worth every cent.

A wood-moving device that may be used within the house is the metal hand truck with reversible handle. They are a tremendous labor saver, but only if the wood supply is nearly on the same plane as the stove; they do not negotiate stairs well. However, they are easy to maneuver and can easily take 75--100 pounds of wood at a time.

You won't forgive yourself if your nicely dried woodpile gets covered with an early six-inch snowfall. Make sure it is covered. Old roofing tin is sturdy and durable, but since many others are also aware of this, suitable quantities are hard to locate. Plastic will do but can deteriorate quickly. A good alternative to either of the above is a rugged canvas cover that has been specially treated to repel water and resist mildew. Grommets should be provided along the edges so that the canvas can be lashed down tightly over the wood pile. A proper-sized canvas tarp can be used to cover other items as well in the off season and will be a useful item for many years.

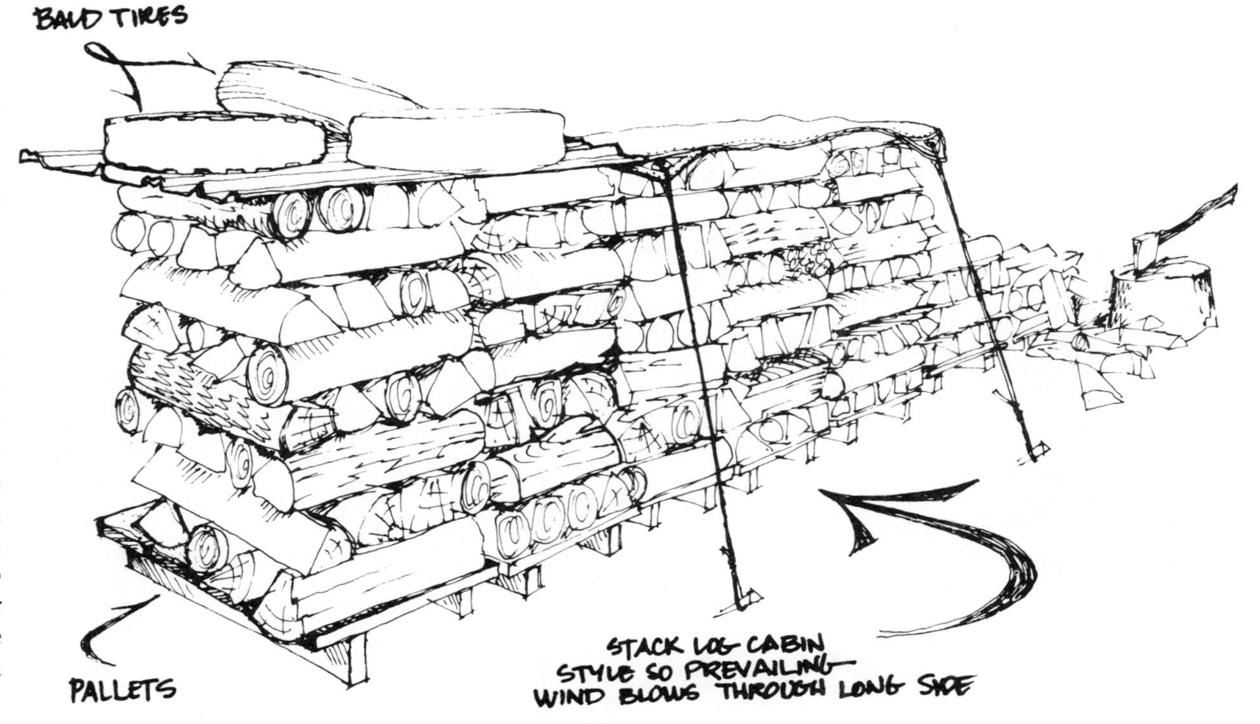

Wood Ashes in the Garden

Once the snow has gone in the spring you will be able to empty your ash storage barrels. These will be twenty or thirty gallon trash cans with lids, or possibly old fifty-five gallon drums. Maybe you have no container, just a large pile of soggy ashes. If this is the case, put containers on your list of items needed for next year. Wood ashes should be used in your garden; they help to complete the natural cycle that burning wood involves.

Woodburners will have quantities of ash that are the answer to a gardener's prayer—a free lime substitute as well as a source of phosphorous and potash (the second two components in the mysterious NPK formula you see on bags of fertilizer). Chemically, lime is the oxide of calcium, Ca_O, which is abundant in wood ash. This is not the same as calcium carbonate, $CaCO_3$, that is slowly available over a period of time. Ca_O, also known as quicklime, is available very rapidly when sufficient water is present. It combines with the water to form the hydrate of calcium, $Ca(O)_2$. This hydrate is readily available for use; calcium ions replace free hydrogen ions in the soil (the culprits causing a low pH or acid soil), and the soil becomes more alkaline, or "sweeter." Obviously, if you live in an area with an alkaline soil, ashes are not for you. Much of the country's soil can benefit from a little sweetener from time to time, though. If you are in doubt, call your county agricultural agent and find out how you can get a soil test done.

On a large scale, use 1000 pounds (one-half ton) of ashes per acre; on a smaller scale, three pounds per 100 square feet. Because most of New England's soil is poor in magnesium, the wood ashes tend to be also. This can create a calcium/magnesium imbalance. To counter this, apply ground magnesium ore, or, for an inorganic but easy solution, one pound of epsom salts for every pound of ashes used. A word of caution here: Because quicklime absorbs water very rapidly, it may dehydrate tender young roots and cause burning. Avoid a heavy application around such plants. Because all the beneficial elements of wood ashes are easily dissolved by water, they are quickly washed out of reach. One application of ashes before planting, and one to two more applications during the growing season works better than a single application in spring. If you have a compost pile, your ashes can be added directly to it, and from there be applied to the garden at any time.

Another caution; do not use ashes in your potato patch. Sweet soil produces scabby potatoes. Watermelons dislike ashes also, and so do any acid-loving shrubs or fruits such as azaleas, ferns, rhododendrons, heathers, blackberries, strawberries, and blueberries. Otherwise, most flowers, shrubs and vegetables will appreciate a yearly wood ash treatment. Other garden uses include tossing ashes through young fruit trees while the dew is still clinging (an old-time disease-fighting measure), and sprinkling ash in circles around young plants to deter cutworms and slugs. There is also a pesky garden fly that lays its eggs slightly below ground level on the stems of broccoli, cauliflower and cabbage plants. The hatched larvae eat the stem for food and seemingly healthy plants may topple unexpectedly. Good success can be had in combatting this by mixing one half cup of ashes per plant into the top inch of soil surrounding the stem when the plant reaches a height of about 10 inches. The flies won't lay their eggs in sweet soil and will search out other nests.

Small amounts of wood ash are relished by all kinds of livestock—from pigs and chickens to horses and cattle. These animals seem to seek out wood ashes, sensing the value to their diet of the trace minerals. Chickens also benefit by having a few handfuls of ash thrown onto their dust bath area. Mites, lice, and other bothersome parasites are controlled by the daily dusting the chickens give themselves.

Learn From the "Experts"

A Tale of Two Ashes
Part 2

"Since I did not have to begin work until noon that day, I used the morning to perform diligent stove-owner tasks, such as cleaning the chimney in my mother-in-law's house. She had left town several days earlier, and the spring weather had not necessitated keeping the stove going.

I brushed the flue and pipe, collecting the creosote in a large paper bag. This task complete, I cleaned the stove as well, adding several shovelsful of "cold" ashes on top of the flaked creosote.

A phone call interrupted my work. I left the bag on the maple floor next to the hearth as I went to answer it. The call occupied me until almost noon, when it was time to head to the job at which I was to spend the next eight hours advising customers on safe and proper stove operating procedures.

At 5:30 p.m. I received a frantic call from my wife reporting that my mother-in-law's house was now engulfed in smoke. By the time I arrived the smoke was so thick that the only way I could enter was by crawling on the floor.

Before long I discovered a basketball-sized hole in the maple floor where the "cold" ashes had been left. I could now see the bag smoldering harmlessly on the basement floor.

My mother-in-law's floor is now repaired, and the only permanent damage was to my reputation among my peers. But I did learn a valuable lesson—"cold" ashes are as harmless as an "unloaded" gun. Each should be treated with equal respect or you risk making an "ash" of yourself and your house."

Bill Busha

Spring Checklist

You've made it through Winter, right? Wrong. This is the season when the rookies and veterans go their separate ways. The rookies forsake their chain saws for the lure of the beach or fishing rod, while the veterans know that this is the best time of the year to get the jump on next Winter.

Cheer up. You've probably built up a roll of fat that needs to be exercised off, anyway. If you are completely organized, the anguished days of Mud Season can leave you with a bulging wood shed or coal bin, and lots of free time to enjoy the summer.

☐ ***Fling yourself into spring with great abandon.*** *One lost day never hurts, so long as you pick up the splitting maul on the next.*

☐ ***Analyze last season's fuel consumption.*** *The information will never be fresher in your mind. Write down your observations about wood types, coal sizes, wood length, and anything else you think noteworthy. You will have a chance to amend your mistakes shortly, and it is amazing how quickly last winter's blunders can fade in the summer sun.*

☐ ***Calculate your fuel savings.*** *Face facts. The most painless way to heat is probably electric until the bill comes. We like our stoves, and we like them particularly because they free us from indentured service to power and oil companies. Your efforts may have saved enough to justify a trip South next winter. Surely, you owe yourself at least a dinner out, compliments of your stove.*

☐ ***Prepare your ashes.*** *Your wood ashes will be going on the garden shortly. As for your coal ashes . . . well, no one has quite figured out the best use for those other than as landfill. Load the kids into the truck, and have a family outing to the dump. You'll see all your friends, and maybe you'll come back with something good.*

☐ ***Check your kindling supply.*** *After months of continuous operation your stove is now in an awkward start-up/shut-down situation. Make it easy on yourself by having plentiful supplies of good, dry kindling.*

☐ ***Order next year's fuel now.*** *Loggers are in the woods, and prices will rise as you get closer to the next heating season. Coal-burners, fill your bins sooner rather than later.*

☐ ***Read the newspaper.*** *This is the leanest time of year for your local stove dealer, and therefore the best time of year to strike a good bargain. If you can anticipate your needs for next season, a new stove at this time of year can be a good way to invest your income tax refund. Buyer beware, however. This is also the time of year when your local stove merchant will be trying his hardest to relieve himself of the unsold dogs in his inventory. Know what you need and get what you want.*

☐ ***Bird-proof your stoves.*** *You would be surprised at how inviting your smelly flue will look to the swallow in search of a nesting place. Some wire mesh or even chicken wire will keep the birds out.*

☐ ***Monitor your flue.*** *This is the easiest time to get lazy about keeping your system clean. Low operating temperatures allow rapid creosote build-up, then hot kindling fires can ignite it. Don't ruin your heating season with an eleventh-hour chimney fire.*

☐ ***Open a window.*** *Your reduced need for heat means that you will be inclined to close your air inlets to the minimum. The result will be rapid creosote build-up and coal fires that simply fade away, leaving you with a very big mess to clean up. More than ever you need to give your stove the chance to operate efficiently. After a lifetime of paying for fossil fuels it is very difficult to adjust to the habit of opening a window so that you can burn your stove hotter and therefore more efficiently.*

☐ ***Burn your junk wood.*** *The small, hot fires that you need at this time of year are perfect for the odd-sized pieces of wood you have been systematically throwing to the side. Also, those punky pieces and scraps of bark will keep you warm so long as you tend the fires carefully. For less heat output while retaining firebox efficiency, burn smaller-sized pieces of wood.*

☐ ***Review backpuffing procedures.*** *Low flue temperatures and gusty winds create ideal conditions for draft reversals. This can be noxious with wood, filling a room with smoke in minutes, but downright dangerous with coal because invisible backpuffs are less detectable. The best defense is a good offense. Burn your stove as hot as possible to maintain flue gas buoyancy.*

2

Summer

Floating Stoves and Cast Iron Boats
Summertime Is . . .
Planning is Half the Fun
Summertime Also Is . . .
Summer Checklist

On the first sunny day after a heavy rain the land turns so green that you need sunglasses to keep down the glare. Daisies explode on the fields only to be devoured in a single day by cows who roam free for the first time since September. This is the season that Vermonters claim offers "damn poor sleddin'."

Floating Stoves and Cast Iron Boats

The stoveowner's world is reversed from the rest of humanity's. Normal people atrophy in the Winter, compensating for the deterioration of the body by feeding the mind with cultural stimulation. When Spring arrives the city dweller casts free his shackles by subjecting his body to the abuse that only intensive lawn mowing, gardening, jogging, and tennis can bring.

The stoveowner, meanwhile, has been flogging his body since September—hauling fuel, emptying ashes, and preparing the next winter's fuel supply. Unless he is reserving the summer for major do-it-yourself projects, by Memorial Day he faces several months when the living is easy and the demands of his stove are nonexistent. It's a good time to think.

Vermont is the only New England state without a coastline, and in the summer the deficiency becomes acute. The Green Mountains are majestic, the sunrises spectacular, and the farms routinely beautiful, but somehow the countryside lacks both the raw force of pounding surf and the salty edge of a clambake. Our landlocked minds fabricate intricate rationalizations to connect our lives with the sea. The results can be seen in Vermont Castings' stoves.

Duncan Syme, the designer, has nautical roots. He grew up on the seacoast and was a professional boat-builder before turning from wood to cast iron. He will also kill for shellfish in any form. His influence is visible in the names of our stoves, which commemorate famous craft that possessed the same attitudes toward the sea as do the stoves to the Winter—Defiant, Vigilant, Resolute, and Intrepid. This same influence can be seen in the whimsical designs, such as the Plimsoll mark that appears on the Resolute fireback to mark the prudent load limit for wood. The mark was originally created to define an acceptable level of submergence for British ships plying the India trade.

The presence of whimsy in stove design is acceptable so long as it is not at the expense of function. Boats and stoves have much in common insofar as their demands on design. Each must achieve a certain level of efficiency and economy within an aesthetic context that allows little flexibility for compromise. The winter and the sea demand this. The finished craft must meet the requirements of beauty and tradition. Their owners, who tend to favor the romantic side of life, demand this. Perhaps nowhere are the rigors of design so apparent as when a stove is installed on a boat.

At first glance, a stove on a boat seems impractical. Cabins are small and permit little room

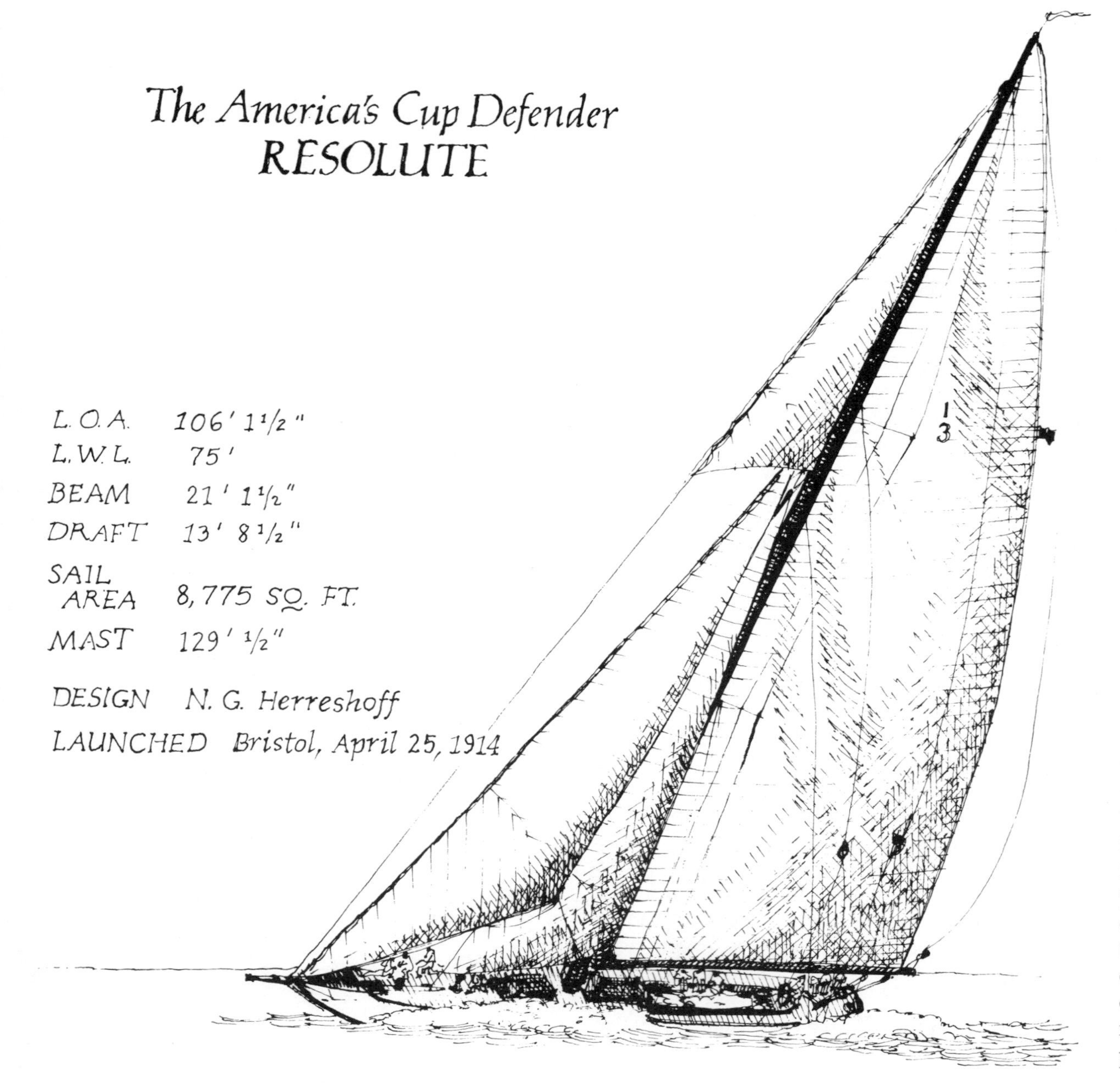

for fuel storage, let alone a stove. The unit must be bolted to the floor, lest it slide at every roll and pitch. The chimney invites trouble beyond the ken of the homeowner. Whenever the boat heels, it is no longer vertical. Not only is draft reduced, but waves can enter through the top. Insurance rates increase. But most importantly, if the weather warrants a stove, then good judgment usually dictates staying at home.

The tradition of the sea and the stove is well-established. Men who work the Northern seas do not retire when winter arrives. Cold-water fishermen continue to work. For years their stoves have served as the most practical method to heat quarters, to cook meals, and to dry bedding or clothing. Luxury yachts often sport elegant stoves and occasionally have open fireplaces.

Nautical installation problems can be solved by working with either a ship chandler, such as Chase & Leavitt in Portland, Maine, or a manufacturer specializing in boat stoves, such as Lunenburg Foundry & Engineering, Ltd. in Lunenburg, Nova Scotia. Lunenburg makes a complete line of boat stoves. Their models include the well-known Sardine, the Little Cod, the Gift, and the Fisherman. The Sardine claims the title of the smallest cast iron stove in the world. It weighs only twenty-seven pounds, uses seven-inch pieces of wood, and measures 12 × 12 × 11 inches. The Little Cod has been on the market for years and is a favorite along the Canadian coast. It can handle 12-inch pieces of wood and weighs thirty-six pounds. The Gift and the Fisherman come in several sizes and burn either wood or coal. The Fisherman is also a cook stove and has a railing around its top surface to keep pots and pans from sliding off. Another nice feature is a heavy oven door that will not fall open when the vessel moves. All of these models have special feet that can be fastened to the floor.

In addition to stoves, Lunenburg carries a line of pipe, chimney caps, chimneys, and deck irons uniquely suited to the marine trade. They make a 17-inch-tall cast iron deck chimney that makes a strong support for pipe. Of particular interest is their line of chimney caps. The Dreadnaught cap has an interesting H-shaped design to prevent water entry but requires more room than a conventional cap. Lunenburg claims that their "O.K." chimney cap model will create an updraft with the wind in any direction, and will shed rain down the outside of the pipe.

Once the stove is installed on a boat, a source of fuel must be found. Coal can be purchased in bags but should be transferred to a waterproof container. Wood is difficult to purchase in small quantities, so it is wise to be creative. Wood scraps can often be found at boatyards. Oak, teak, and even mahogany scraps can sometimes be had at very reasonable prices. In fresh water, driftwood can be a good fuel source, but salt-water driftwood will corrode the pipe. The first-time stove installer must remember that boats do not remain level, so a woodbox or coal bin is helpful to keep a mess off the deck. It is also wise to keep stove doors tightly secured so the stove's contents remain there.

Once accomplished, the union of stove and boat provides far more than heat. It is a marriage that allows one to remain comfortably afloat long after stoveless craft are mothballed in winter storage. The tradition continues. But most importantly, a special need is filled, a need understood only by those drawn to the sea.

A Boat Stove

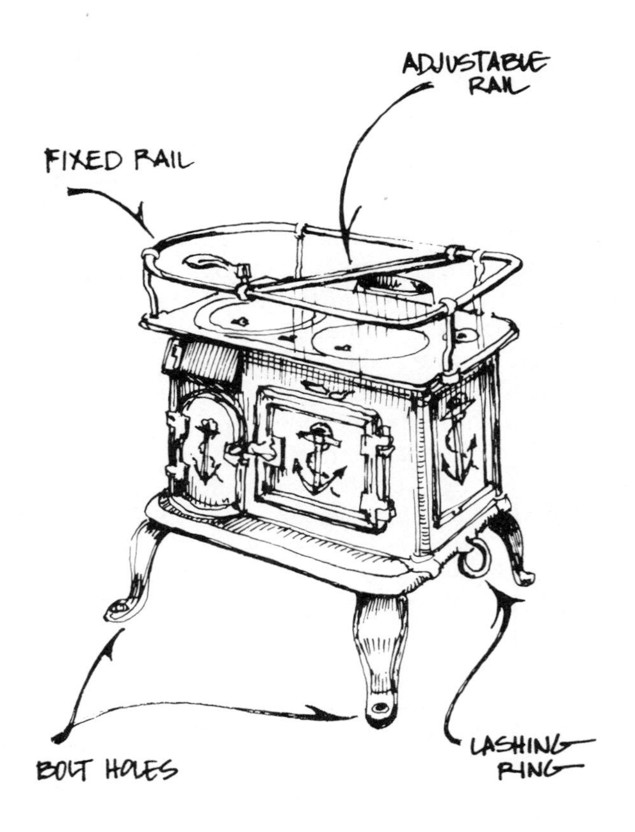

Dreadnaught Head

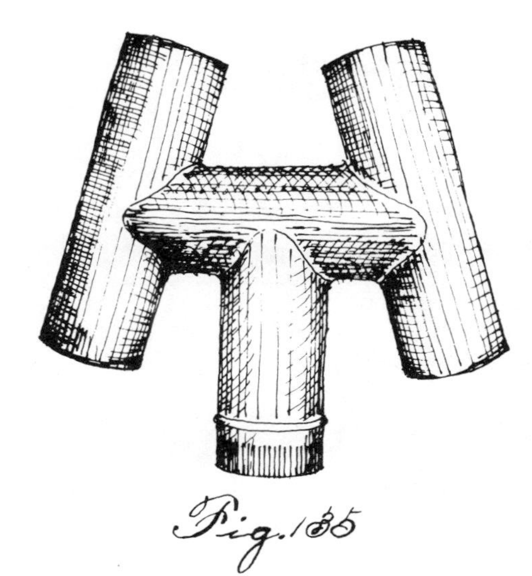

Summertime Is . . .

The Atlantic Ocean, even at high tide, is more than a hundred miles from the sloping foothills of Vermont's eastern border. Although the challenge of the sea remains at a comfortable distance, Vermonters nurture their own reverence for the demands of the elements and the seasonal challenges with which they cope. For example, one tired joke that travels through the state characterizes the Vermont year as "nine months of winter and three months of damn poor sledding." The joke probably started sometime after 1816, a year that brought Vermonters a killing frost in each of the twelve months. Fortunately, that is not typical Vermont weather, though, and we can usually depend on four full months when our stoves do not need to be tended.

As in hundreds of communities everywhere, the mild summer months in Vermont are the ones during which projects hopefully are accomplished. Gardens are planted, houses are painted, and garages cleaned out. The industrious stove-tender will do these things too, but there are other projects for which he has spent the winter making plans as well. Some of these projects will never come to fruition, like the forty-foot woodshed that was to be built on his mate's flower beds. Other projects are essential, like building a safe hearthpad, and must not be victimized by procrastination.

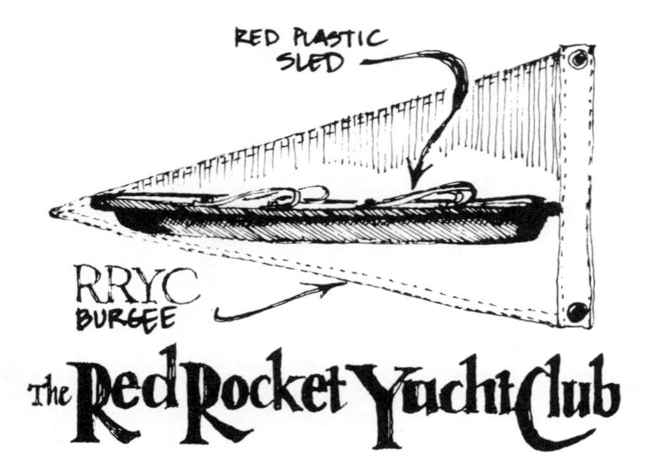

On the next few pages are described typical summer projects that will enlighten your heating routine. You do not have to be a master carpenter to complete the projects; they are ideal for those of you whose ambition looms larger than your skill. Each item will be thoroughly functional in its finished form and can be as decorative as your talent and imagination allow.

A Solar Dryer

Building a solar dryer can be very simple. Pile your wood in a sunny spot on logs or pallets; cover the pile with plastic, make a space at the bottom and a couple of holes at the top for air flow, and you've got a solar dryer. There are many possible designs. They can be small or large, sophisticated or simple, permanent or temporary; they all work.

We have found the wood dryer used in tests at Vermont Castings to be very satisfactory. We designed it to be simple to construct, inexpensive, and highly effective. It can be put up in an afternoon using a few basic tools. The materials are available at any building supply store, and the total cost is around $30.00. It is built in separate panels that are bolted together and can be taken apart and stored flat when not in use.

For a temporary structure we built the dryer using strapping and 6-mil plastic. For a more permanent structure which can double as a woodshed, we recommend using 2- × 4-inch lumber with corrugated fiberglass for the roof and some extra structural pieces. This heavier dryer will cost about $85.00 to build.

The dryer measures nine feet long by five feet wide and holds over a cord of wood. To dry more wood at one time, you can make it as long as you wish, in nine-foot units.

Hearthpads

A story in the local weekly newspaper recounted the details of a house fire. The woodstove had been installed directly on the bare floorboards with no hearth protection underneath and no heatshield mounted to the stove. As will happen, the unprotected floorboards dried and darkened over a period of time and, on this particular day while the occupants of the house were away, caught on fire. The boards under the stove charred to such an extent that they could no longer support the stove, and the hot unit plummeted into the basement below. Fortunately, the stove severed a water pipe in the course of its fall and the ensuing spray was enough to extinguish the flames before any further damage was done.

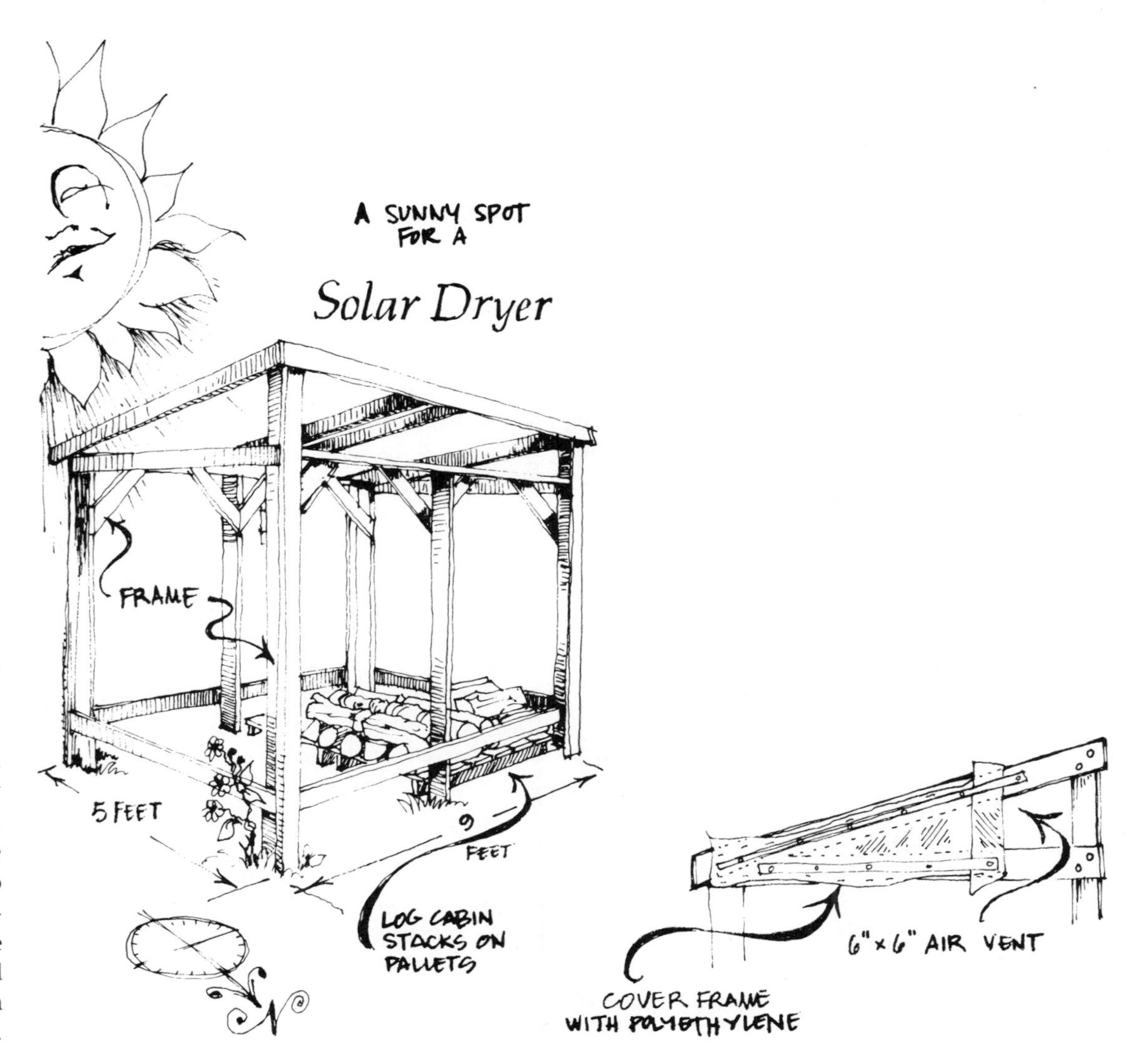

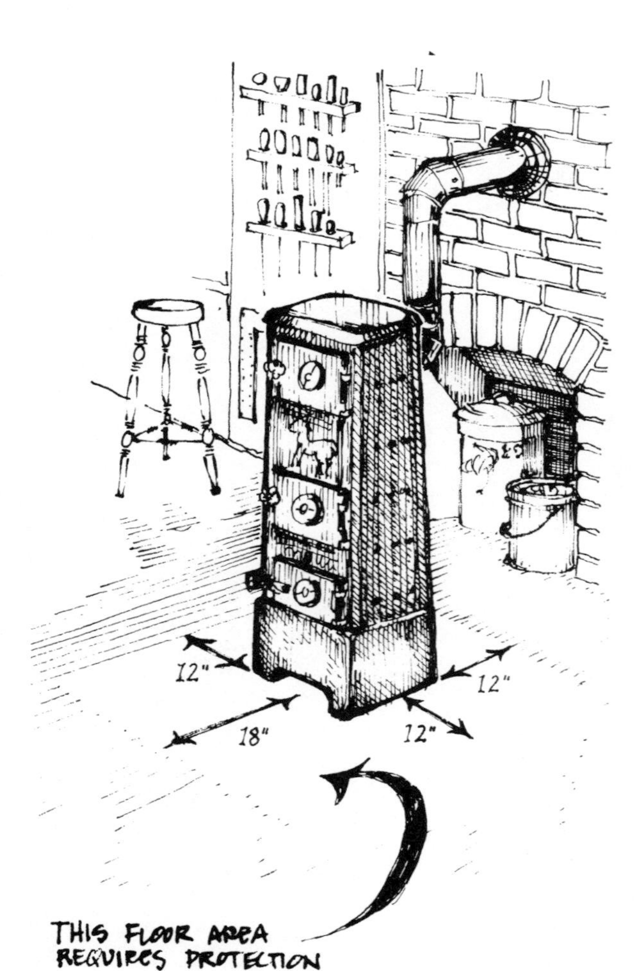

Although extremely unusual, this situation actually happened in central Vermont and could happen again in an installation that does not use a suitable hearthpad.

In the proverbial olden days, hearthpads were seldom used; if they were part of the installation at all, they would not have been acceptable by the more rigid safety standards that exist today. Back then, the hearthpad may not have been as critical to the safety of the installation as it is now. Old-time parlor and cookstoves were basically updraft stoves. The heat they generated traveled in an upward direction, through the stove top and into the stovepipe. This combustion characteristic was also responsible for the low level of efficiency of those stoves; much of the heat was lost up the flue.

Contemporary airtight stoves, on the other hand, are designed to keep the heat within the body of the stove as much as possible in order to minimize heat loss in the stack. This results in the stove plates, particularly the bottom ones, staying hotter for longer periods of time than the earlier heating counterparts.

The need for adequate floor protection is essential in any installation, and this protection is provided with a hearthpad.

Build Your Own Hearthpad

Attractive and functional hearthpads may be purchased fully-assembled for $200–$400, but many people prefer to construct their own at a fraction of the cost. In either case, the basic materials are the same: a base of ½-inch plywood, two sheets of mineral board, and an aesthetically pleasing and noncombustible covering such as slate or brick. The entire pad is framed in a hardwood border.

Begin the construction of your hearthpad by determining how large it should be for your stove model. The pad should be large enough to extend twelve inches beyond all sides of the stove. If a particular side includes a loading door, the extension should be for eighteen inches rather than twelve. Once you have determined the proper size, cut your sheet of plywood to fit. Cut the mineral board to cover the plywood as well. Be sure to wear an appropriate air filtering mask. Mount the hardwood border around the perimeter at this time.

Cement the first sheet of mineral board to the plywood with a non-flammable glue, and fasten the second sheet directly to the first. The slate or brick covering should be cemented into place using a commercial mortar mix. The individual pieces should be mortared together, and to the base beneath. A two-gallon quantity of cement is about right for the construction of one pad. The slate may be sealed with a non-flammable sealer to complete the job. Let the materials harden for two days before placing the stove on top.

A Woodbox Bench

A solid woodbox located conveniently near your stove will improve your life in several ways. It will help keep the floor clean and unmarked and thereby improve relations with your housemate. It will provide a constant supply of wood next to the stove so that trips to the woodpile can be made at your convenience rather than out of desperation and, if equipped with a closing lid, it will give your cat a warm place to sleep, freeing up the favorite chair for you.

Your woodbox may be as crude or elaborate as you want it to be. In the summer it may be moved to the porch and used for a bench and for storage.

A woodbox, as with any combustible material, should be at least 36 inches from the stove. If it is necessary to place it closer, be sure to adequately protect it with an appropriate heatshield. A heatshield may be made from a piece of sheet metal. Attach it so that a one-inch air space exists between the shield and the box.

A Coal Bin

Proper size, construction, and placement of your coal bin is important in order for coal to be a convenient fuel.

The size of the coal bin really depends on how much coal your furnace or stove uses per day, and how easy it is to purchase and receive delivery of the coal. Generally, coal stoves will consume from 45 to 100 pounds per day. This means storing between one and one and a half tons if you expect a delivery each month, or two to three tons if you expect a delivery every couple of months. If taking delivery is a problem, or if coal is hard to find and you have the space, you may want to invest in a larger bin. Anthracite coal generally fills about 37 cubic feet per ton. That is a

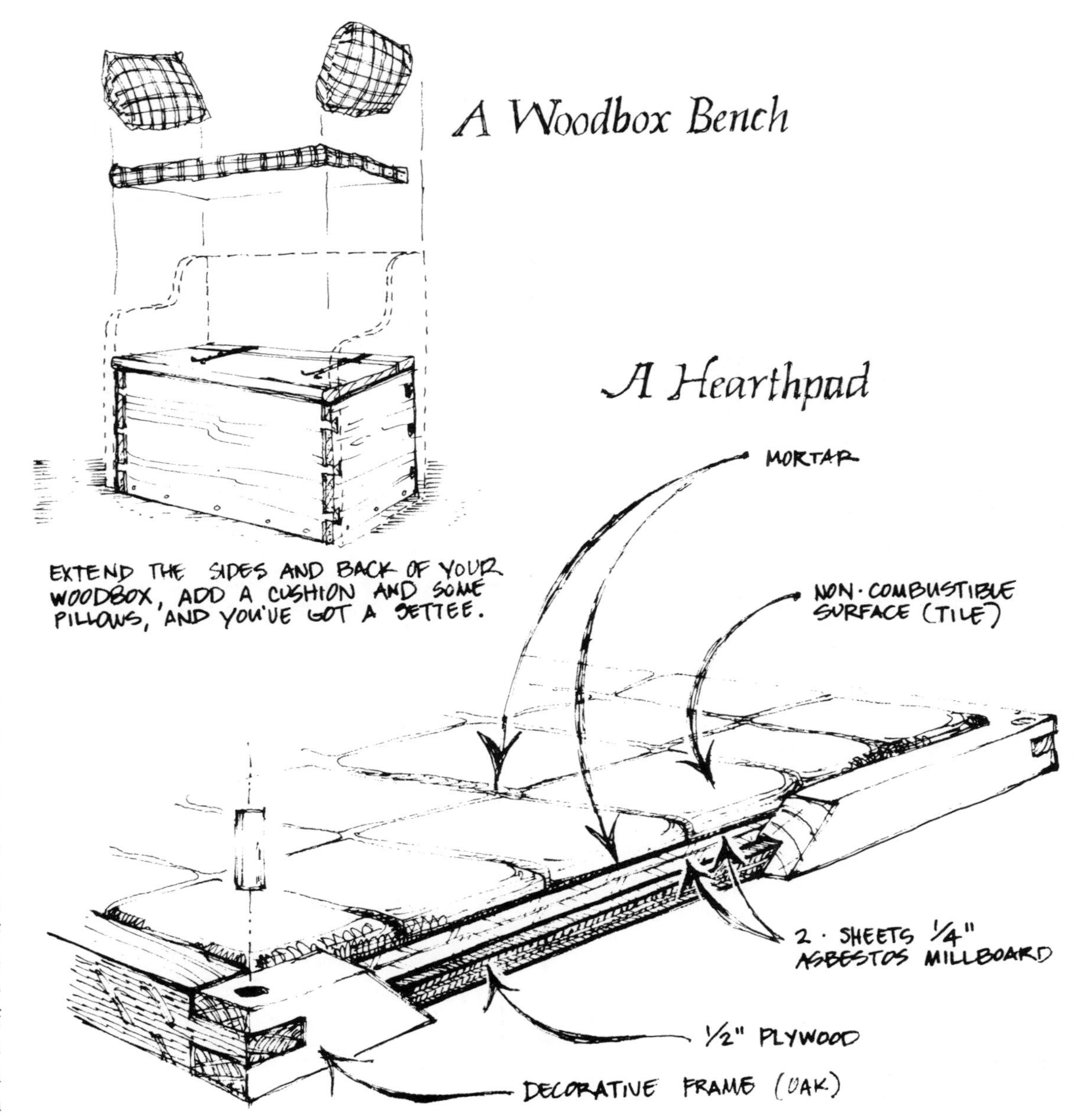

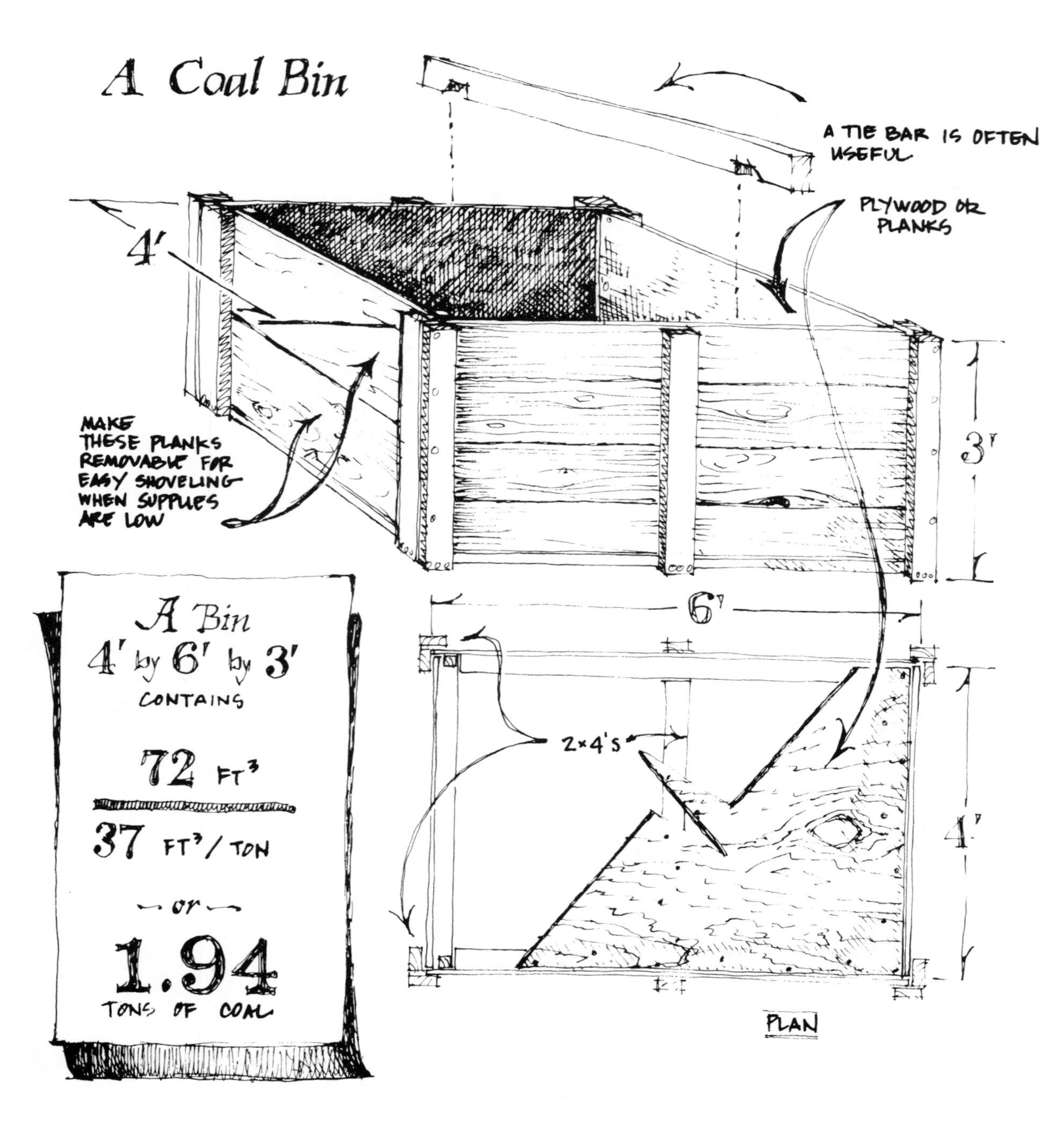

bin 3 × 3 × 4 feet. Bituminous coal takes a somewhat larger space at 44 cubic feet per ton.

The dense nature of coal makes it important to build a sturdy bin. Pouring coal into a poorly constructed coal bin can be a lot like pouring concrete into a weak form. First the sides will bow outward, then a side will break, then all of a sudden the coal is flowing everywhere. Use one-inch boards or plywood for the sides of the bin and reinforce it well with 2- × 4-inch lumber. On long bins be sure to tie the top edges together, or your bin will end up looking more like a boat than a bin.

A proper location for the bin is as important to convenient coal burning as having a well-designed "convenience" triangle between your cooking stove, refrigerator, and sink in a well-designed kitchen. Construct the bin in a place that makes it easy to transfer the coal from the delivery truck to the bin and then to your stove or furnace.

Planning is Half the Fun

A major summer project that precedes the suggestions of the previous pages is the actual installation of a stove. It may be your first installation, or it may be the second or third one within the same house. A thorough evaluation of your plans for the stove will help you to get maximum satisfaction from it. One of your first decisions will be the type of fuel you will burn.

Wood is man's original fuel and is still the least expensive form of heat in many parts of the country. The main disadvantage of wood is the difficulty and expense of transporting it. In areas where these problems make wood an impractical alternative, coal may be the answer. Coal is also an inexpensive form of heat and has the major advantage of having a minimal handling requirement; it can usually be dumped in your garage or basement. Anthracite is the only coal that you should consider burning, as bituminous can create undesirable environmental problems.

The best solution to the fuel question is to buy a stove that will burn either wood or coal. You will then have the versatility to burn whatever is least expensive and most plentiful.

A second decision will need to be made about how you plan to use your stove as a heat source and where in the house you are going to put it. There are several factors that will influence the placement of your stove, but first consider the part your stove will play in the overall heating plan for your home. Most stoveowners prefer to use a stove as a *primary* source of heat. A primary source supplies 50% to 75% of the heat needed with a back-up source of heat for times when the winter temperatures are at their coldest or when the stove-tenders are away for a weekend or longer.

Others prefer to use their wood or coal stoves as a *secondary* source of heat to supplement the main heat source (oil, gas, or electricity) or to provide heat for a part of the home that the main source cannot reach. A third option is to rely on your stove as the *sole* source of heat for your entire home. Generally, homes that employ a parlor stove as the sole source of heat are very well-insulated. Your ultimate decision on where to locate the stove will depend on how much you plan to rely on the stove as a heat source.

Be realistic about your goals, and be flexible in how you will meet them. Don't plan to heat your eleven-room house evenly with one parlor stove when the weather is below zero. Let your stove be the sole source of heat for most of the winter, but don't be too proud to turn on the oil

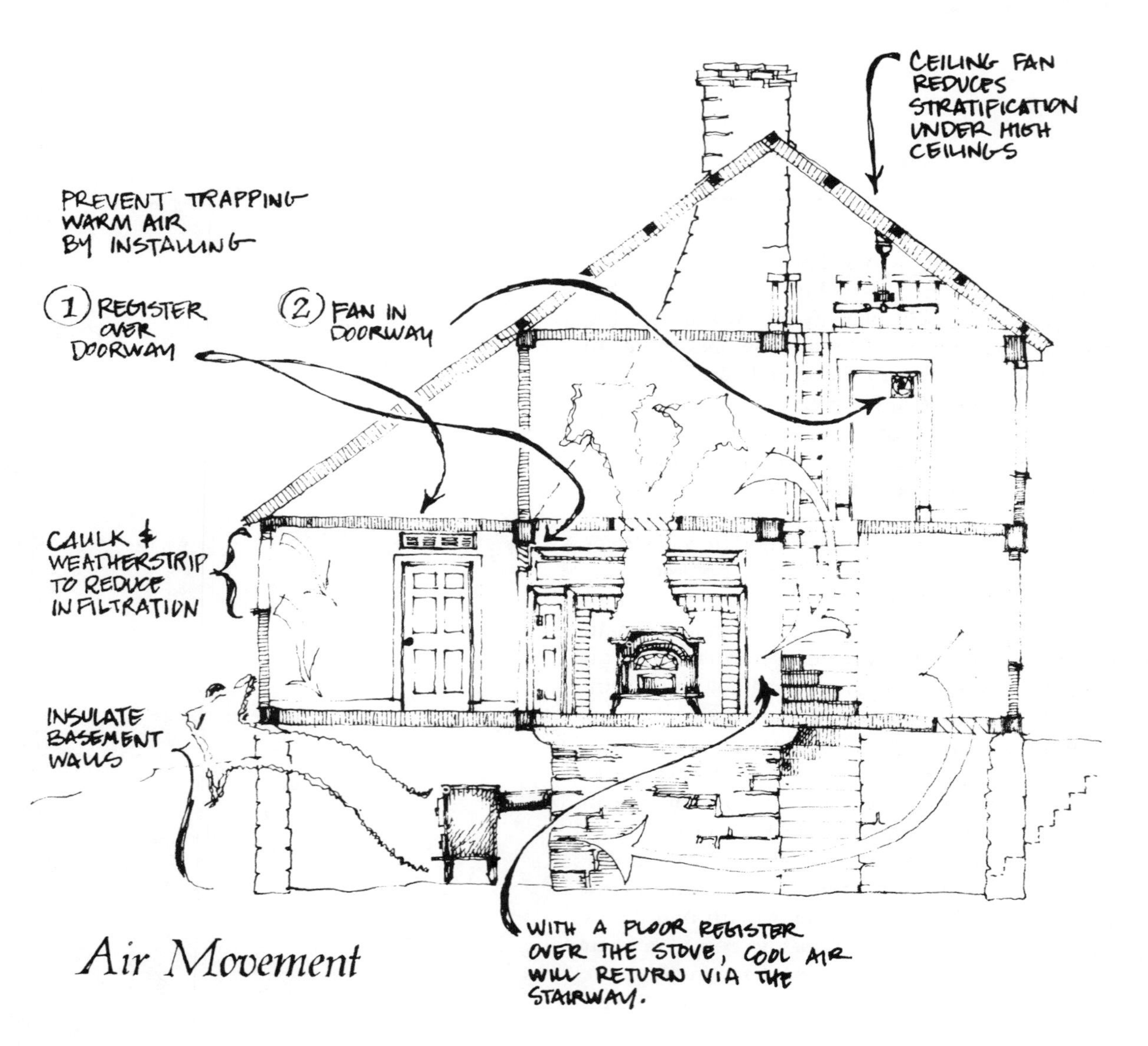

Air Movement

burner or gas furnace for that one-week cold spell. Location of the stove within the house will have a great influence on how effectively the heat is utilized.

Basement installations are most effective for heating the basement and, to a lesser extent, the floor above. Floor grates will facilitate the passage of warm air to the first floor, and the stairway is a natural avenue for the heat as well. Basement walls should be insulated to prevent heat from being absorbed by their cool surface. It is often difficult to achieve satisfactory and uniform whole-house heating from a basement installation. First-floor installations usually provide the best all-around heating potential.

The best heating results will usually be achieved when the stove is located in or near the center of the house. Distant corners will be cooler than the area directly around the stove, but this offers a range of temperatures to suit the comfort preferences of different family members and for different activities.

A stove located in a corner or against a wall of the house will provide comfortable heat for the main living area surrounding the stove. The far end of the house may be 10 to 15 degrees cooler. In many homes the cooler end is also the sleeping area, and the cooler temperature range is desirable.

Most homes contain architectural features that influence the level to which they may be heated efficiently. An open stairway, for example, will allow warm air to rise readily to the floor above. Stoves placed very near the stairway may allow more heat than is desirable to rise directly upward. In order to force more lateral movement of the warm air through the house before it rises, the stove should be located at a greater distance from the stairs.

High ceilings can allow the warm air to rise and stratify in the top few feet of the room. This results in a great deal of the heat pooling in an area where it cannot be used, while allowing the first five or six feet off the floor to remain uncomfortably cool.

Older houses often are constructed with smaller and greater numbers of rooms per floor than more modern constructions. Walls are barriers to air flow as well, and some benefit can usually be realized with the installation of wall registers. Another option in houses that are divided into many compartments is to use two smaller stoves on opposite ends of the house instead of a single large stove.

Fans and Air Movement

Many of the factors that impede air movement can be improved through the use of fans. A strategically-placed fan can assist the natural air flow patterns within the house or can artificially create an air flow cycle where none previously existed. Fans come in many shapes, sizes, and designs, and many are designed for a specific purpose. Consider the following criteria before buying a fan:

- The fan should encourage air movement.
- The fan must operate economically. Since one of our objectives in owning a wood or coal stove is to save money, it doesn't make any sense to buy a fan with high operating costs.
- The fan must be designed for continuous operation. The proper fan to use for heat circulation is one that moves air continuously to maintain even heat distribution.
- The fan must be extremely quiet.

The first type to consider is a small fan to warm up a single room or small area. These fans were originally designed for use in cooling computers and are engineered for continuous operation. The size of these fans is ideal as they are small enough to hang from a doorway or be placed on the floor and not be an obstruction. They will move about 60 cubic feet of air per minute (60 CFM) continuously and quietly.

A second type of fan is designed to be installed in a wall with grates on each side. This allows you to remove the effects of a wall heat barrier and move air directly from one room to another. These fans are available from several manufacturers and are fairly similar in design. They are designed to fit between wall studs and are often adjustable for wall thickness. They usually have speed controls and can move up to 200–225 CFM. They use about the same amount of electricity as a 40-watt bulb and are relatively quiet, especially at low speeds.

One point to consider before buying this type of fan is that they create a permanent opening between two rooms and allow sound transfer between the rooms. It is wise to hire a licensed electrician to install the fan in the wall.

A third type, the ceiling fan, or paddle fan, the popular old standard of the South, has been revived to combat energy costs. the original ceiling fans go back at least 60 years. They were used for cooling homes in the South before air conditioning. Now that energy costs are so high, ceiling fans are being used for both heating and cooling. On high speed they are very effective for cooling purposes; on low speed they pull warm air down from the ceiling without creating a cooling breeze. This allows you to keep rooms with ceiling fans at a lower temperature and still be comfortable. Ceiling fans are most effective in homes with high ceilings. They cannot be used on ceilings under 8 feet for obvious reasons. Ceiling fans are available in a wide array of styles, from very plain to extremely ornate, from painted to brass plated. Prices vary from about $100 to more than $1,000.

There is a wide variety of accessories available for the more expensive ($200 and up) fans including trim packages, light kits, and blade options. Most ceiling fans have about a 52-inch diameter, with a 36-inch model available for smaller rooms. Ceiling fans are not only useful for reducing air stratification in winter and cooling in summer, but are also gaining popularity as decorating pieces.

Fans

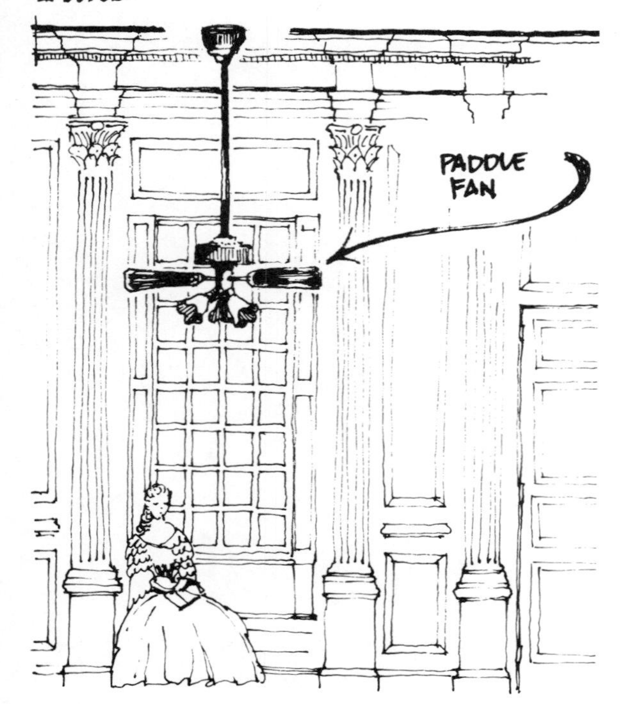

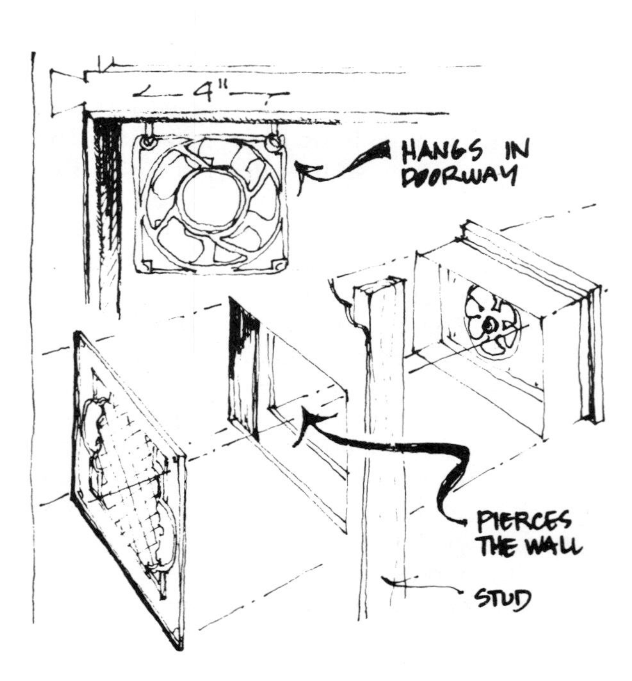

Your heating expectations, the placement of the stove, and your ability to use the heat effectively will influence your level of comfort. Equally important is the efficiency level of your existing house. The tightness of construction, level of insulation, and presence or absence of heatsaving devices such as insulating window shades are all factors to consider. Efficient heating involves retaining heat in addition to generating and circulating it.

Clearance Requirements— How Far Is Too Close?

In the preface of his book, *Wood Heat Safety,* Jay Shelton poses the question, "Is heating with wood (and with coal) safe?" His answer, "Probably . . . if . . . ,"* exemplifies the complexities that

*Jay Shelton, *Wood Heat Safety* (Charlotte, VT: Garden Way Publishing Co., 1979), p. ix–x.

confront anyone wishing routinely to build a lively fire within his home. Accomplishing this feat in a manner that allows you, your family, and your neighbors to sleep securely requires that two basic problems be addressed simultaneously: 1) how to keep the fire contained within the heating system and 2) how to prevent spontaneous combustion of surrounding materials. Assuming that the components of the heating system are in good condition and are properly constructed or installed, the first problem is easily solved through correct operation and maintenance. The second problem, somewhat less obvious than the former, but no less severe, can be solved in a number of ways that we will discuss here.

The term "spontaneous combustion" conjures up visions of a burst of flame—swift, sudden, and uncontrolled. In fact, however, wood can burn only after it has undergone a series of chemical reactions resulting from exposure to heat. This process, known as *pyrolysis,* can be an insidious precursor of doom to the stoveowner who chooses to skimp on clearance precautions. When the wood is heated, chemical compounds are produced that form gases that are then driven off. As this thermal degradation progresses, the surface gradually darkens, and, if the process continues unchecked, the wood will eventually blacken, becoming charcoal. Since charcoal can ignite at temperatures as low as 300°F, the possibility of such an occurrence taking place within the proximity of a radiant heater is quite real. Depending on the intensity of heat and extent of exposure time, the entire transformation can take place within a period of hours or a span of years.

Some time ago, we were treated to an unexpected demonstration of pyrolytic action during a major renovation of our Randolph showroom. When the oak floor was cleared of stoves and hearth pads in preparation for refinishing, a telltale rectangle of slightly darkened wood marked the spot where an operating Defiant had resided for five years. The hearth pad, obviously, had not been performing its protective duties as well as intended. A quick investigation revealed that construction materials in the pad consisted of ½-inch quarry tile mortared over ¾-inch plywood—a combination that is adequate only when a bottom heat shield is mounted under the stove. For five years the bottom plate of the Defiant had been steadily radiating heat to the hearth pad which conducted the heat through to the wooden floor. Although this situation posed no immediate threat, the odds that the floor would eventually ignite were getting better with each passing day.

No doubt, variations of this somewhat embarrassing tale are being played out in hundreds (thousands?) of homes throughout the country. Studies conducted by government, insurance, and fire-prevention organizations conclude again and again that unsafe installations are the cause of the overwhelming majority of stove-related house fires. And improper clearances are the most common facet of an unsafe installation. Unfortunately, it is all too easy to compromise on clearance dimensions of shielding materials for the sake of convenience or economy. The resultant hazards are not obviously apparent.

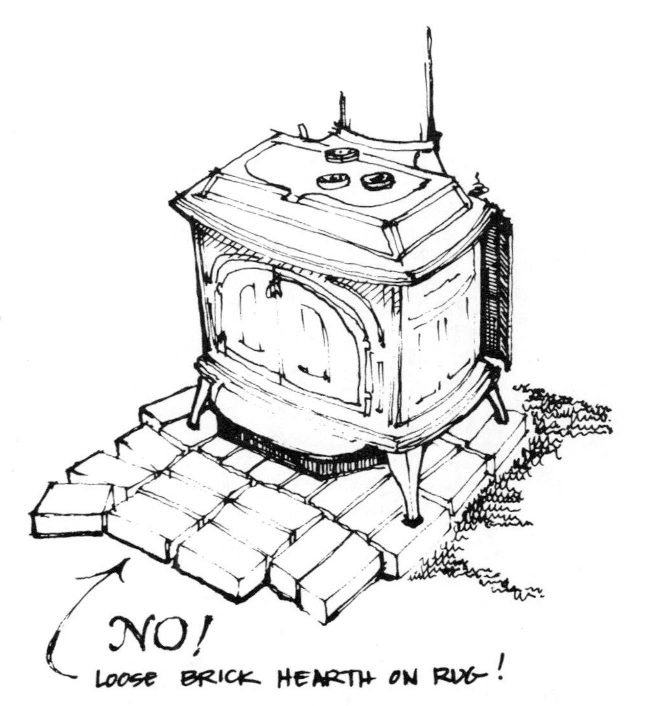

Who Decides?

So, just how far is too close? As you might expect, the answer depends on a wide variety of factors relative to the installation type, test results on the specific stove, and local building codes. While the stove manufacturer, testing laboratories, and independent code organizations all publish their own clearance specifications, ultimately it is only your own local building inspector who can issue an official proclamation on a particular clearance requirement. Building officials, however, generally rely on the recommendations of nationally recognized independent testing agencies for data upon which to base their decisions. There are several such agencies, and sometimes there are strong regional preferences for one over the other. Four of the most common are Underwriters Laboratory (UL), International Congress of Building Officials (ICBO), Building Officials and Code Administrators, (BOCA) and Southern Building Code Congress (SBCC).

When a particular stove has not been tested by an agency, building inspectors will make certain that the installation conforms to local codes that are generally based upon the recommendations of the National Fire Protection Association (NFPA). Their standard #211, which has been continually revised since 1906, includes basic guidelines that are designed to insure the safety of any stove installation operating under the most extreme conditions; i.e., high firing rates for extended time periods. NFPA clearances are not tailored to specific stove types or sizes.

In 1979, Underwriters Laboratory issued the first edition of the *Standard for Solid Fuel Room Heaters,* #1482. This safety standard has since been adopted by building code organizations throughout the nation as the basis upon which wood and coal stoves can be judged to be safely installed and operated. When a wood or coal stove is "listed" by a nationally recognized testing laboratory or regulatory agency, it has passed

a series of stringent safety tests specified by UL 1482. In part, these tests establish adequate clearance requirements for that particular heater. The clearances, indicated in the manufacturer's instructions and on a metal label attached to the stove, supersede the NFPA recommendations. The stove can be installed in accordance with the listed clearances except in the case of conflicting local building codes.

You should also clear a new installation with your insurance company, although they will probably have the fewest specific demands. Often, the insurance company only asks to be notified when the unit is installed; they are not concerned with the details of the installation but want to be informed that the stove is there and installed according to someone else's official recommendations. Some companies, though, will be more stringent. They may require that the installation be inspected by one of their representatives and that a signed inspection certificate be issued before coverage can be guaranteed. Other companies may not cover *any* solid fuel device in some structures without special equipment; the mobile home is one type of structure that may not qualify for coverage once a stove is installed. In any case, check with your insurance company before proceeding with a new installation.

Safety Testing Procedures

All Vermont Castings' wood stoves and coal stoves are tested to be in accordance with UL 1482. The testing programs, conducted by our Research and Development department and supervised by independent testing laboratories,

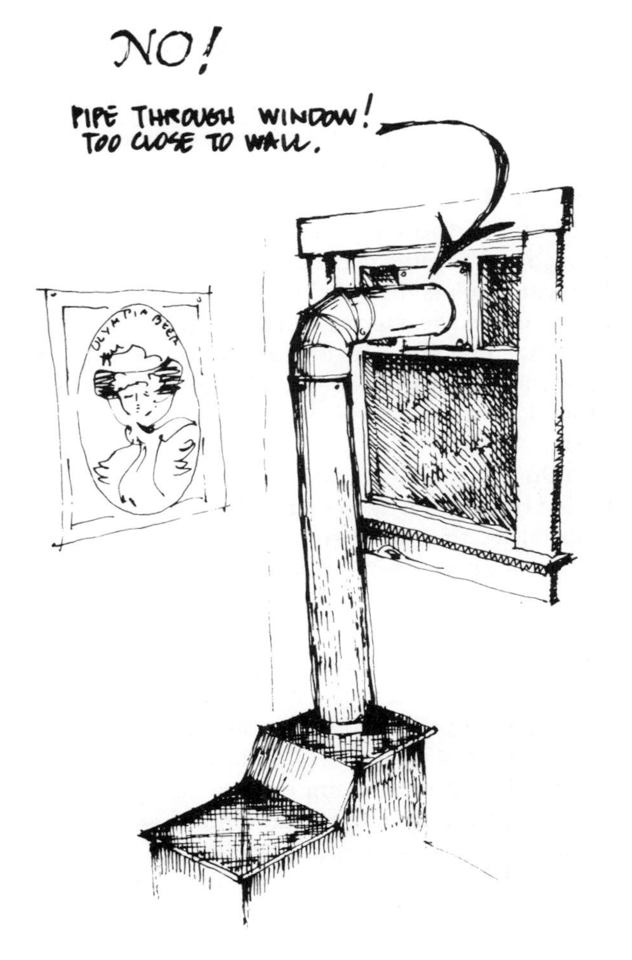

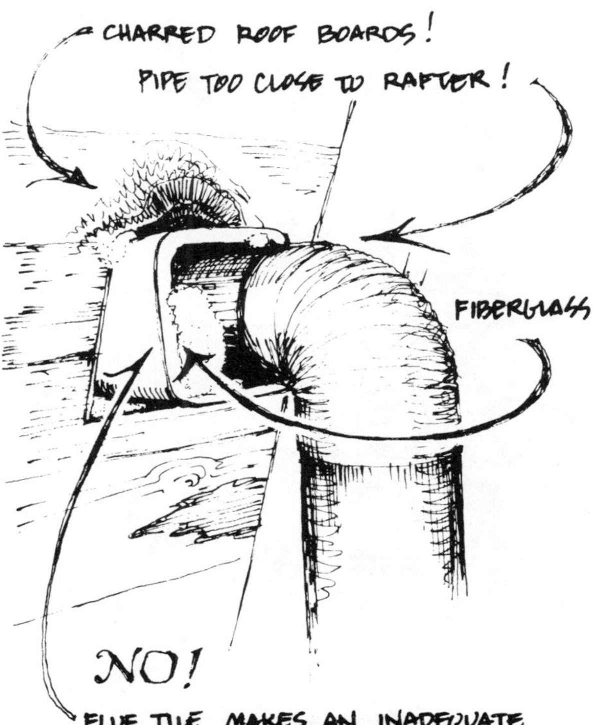

establish specific clearance allowances for each particular stove under a variety of stringently controlled operating conditions. The stoves are installed in a test booth, a combustible wooden structure consisting of two walls, a ceiling, and a floor, and constructed to UL specifications. The walls are portable to allow quick clearance changes without having to move the stove and adjust the chimney connection. During a given test period, temperatures are monitored by as many as 113 thermocouples (heat-measuring sensors) secured to specified locations on the stove and test booth surfaces. Clearances are determined through three separate operational tests: a radiant fire test, a brand fire test, and flash fire test. Each is intended to subject the stove and test booth to the effects of an intensely hot fire and simulate the most severe conditions as can be reasonably controlled. All air controls are removed, and the stoves are allowed to run wide open.

During the radiant test, an extremely hot fire fueled by a 6-inch deep bed of charcoal briquets is maintained until temperatures within the test enclosure stop climbing. Each thermocouple location is monitored every 30 minutes for a period of several hours. Ambient (surrounding) room temperature is also recorded simultaneously. In order to pass the test, surface temperatures of exposed materials (unprotected by a wall shield or hearth pad), may not exceed 117°F above the ambient temperature. Surface temperatures of a protected floor or wall may not exceed 90°F plus ambient temperature. For example, if the maximum room temperature is 78°F during the test period, the surface temperature of an exposed wall behind or beside the stove may not exceed 117°F + 78°F, or 195°F. If the wall temperature climbs higher than 195°F, clearance between the stove and wall must be increased and temperatures again monitored. When the wall temperature falls within the prescribed limit, the distance between wall and stove can be said to be the minimum clearance for that particular stove in that particular installation configuration. In this way, established clearances take into consideration the radiant properties and maximum heat output of stoves of differing sizes and constructions. Consequently, small stoves sometimes qualify for closer placement to combustible materials than larger stoves.

The brand test utilizes fir fire brands of uniform size and moisture content to generate a very hot fire for as long as is necessary to reach maximum temperatures within the test booth. Air controls are removed and the doors left open. Temperatures of the various test enclosure locations must remain within the same limits as in the radiant test.

The flash fire test, designed to simulate out-of-control conditions, can be quite dramatic. Again, fir fire brands are used as fuel, but are loaded into the stove eight at a time rather than singly. The resulting "blast furnace" effect generates temperatures far in excess of any likely to be encountered through normal stove operation. Test booth surfaces, monitored every five minutes until maximum temperatures are reached, must not exceed 140°F over ambient air temperature.

The clearance requirements established for Vermont Castings' stoves as a result of these tests are detailed in Table 2–1. These are the minimum dimensions that can be maintained safely between the stoves and their chimney connection and any adjacent combustible material. These specifications should not be applied to stoves of other manufacturers. Most reputable stove manufacturers will have had their product tested to UL standards, and the stoves should be installed in accordance with those test results.

In any case, the person who builds a lively fire within the confines of his combustible castle will do well to remember that safety is a relative thing; one man's security may be another man's belly-laugh. And we are, after all, dealing with a bit of magic here. Remember the "Lesson of the Showroom Floor." When it comes to putting space between that which is combusting and that which may combust, more is always better.

The clearance that is most often violated out of ignorance is the area around the thimble. As a result, fire investigators claim that the thimble area is the location where a large percentage of fires begin.

If your installation must pass through a wall, make sure there is nothing combustible within 18 inches of the pipe all around. If the pipe is eight-inch, your hole in the wall will be 44 inches square. The hole will be 42 inches for six-inch pipe. The hole is a big one, but not nearly as large as the one that would be created if a fire is caused by inadequate clearance.

Table 2–1 Clearance Table for Vermont Castings Stoves

		Defiant/Vigilant			*Resolute*			*Intrepid*		
		Heat Shields			*Heat Shields*			*Heat Shields*		
		None	*Rear Only*	*Rear & Stovepipe*	*None*	*Rear Only*	*Rear & Stovepipe*	*None*	*Rear Only*	*Rear & Stovepipe*
Unprotected Wall	Side	36	36	36	24	24	24	24	24	24
	Rear[3]	36	25/10	10	36	25/10	10	30	30/16	16*
	Corner[2] Installation	36	18	18[4]	30	30	12	30	30	30*
Protected[1] Wall	Side	14	14	14	8	8	8	12	12	12
	Rear	18[6]	18[6]	6	10	8	6	16	16	9*
	Corner[2] Installation	18[7]	10[7]	10[5]	12	12	6	10	10	10*

Notes:

1. Protected Wall: 1/4" cementboard, asbestos millboard, or noncombustible mineral board spaced one inch from combustible wall on noncombustible spacers, or equivalent. Walls are considered combustible if any part of the wall will burn. Noncombustible materials glued to sheetrock or plaster over wood is not considered adequate wall protection.
2. Corner installation: clearances are measured from rear corners of stove to wall.
3. Upper figure is for top-exiting stoves. Lower figures is for rear-exiting stoves.
4. Stovepipe heat shield not required.
5. Stovepipe heat shield not required for top-exit Defiant only.
6. Vigilant—10"
7. Vigilant—14"

**Stovepipe heat shields must not extend to ceiling. For reduced clearances behind top-exiting Intrepid installed so that pipe extends vertically into prefabricated chimney, use wall protection as specified in Vermont Castings Installation Planning Guide.*

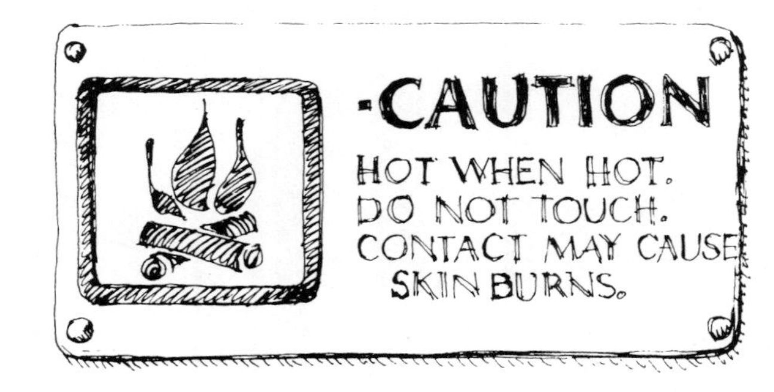

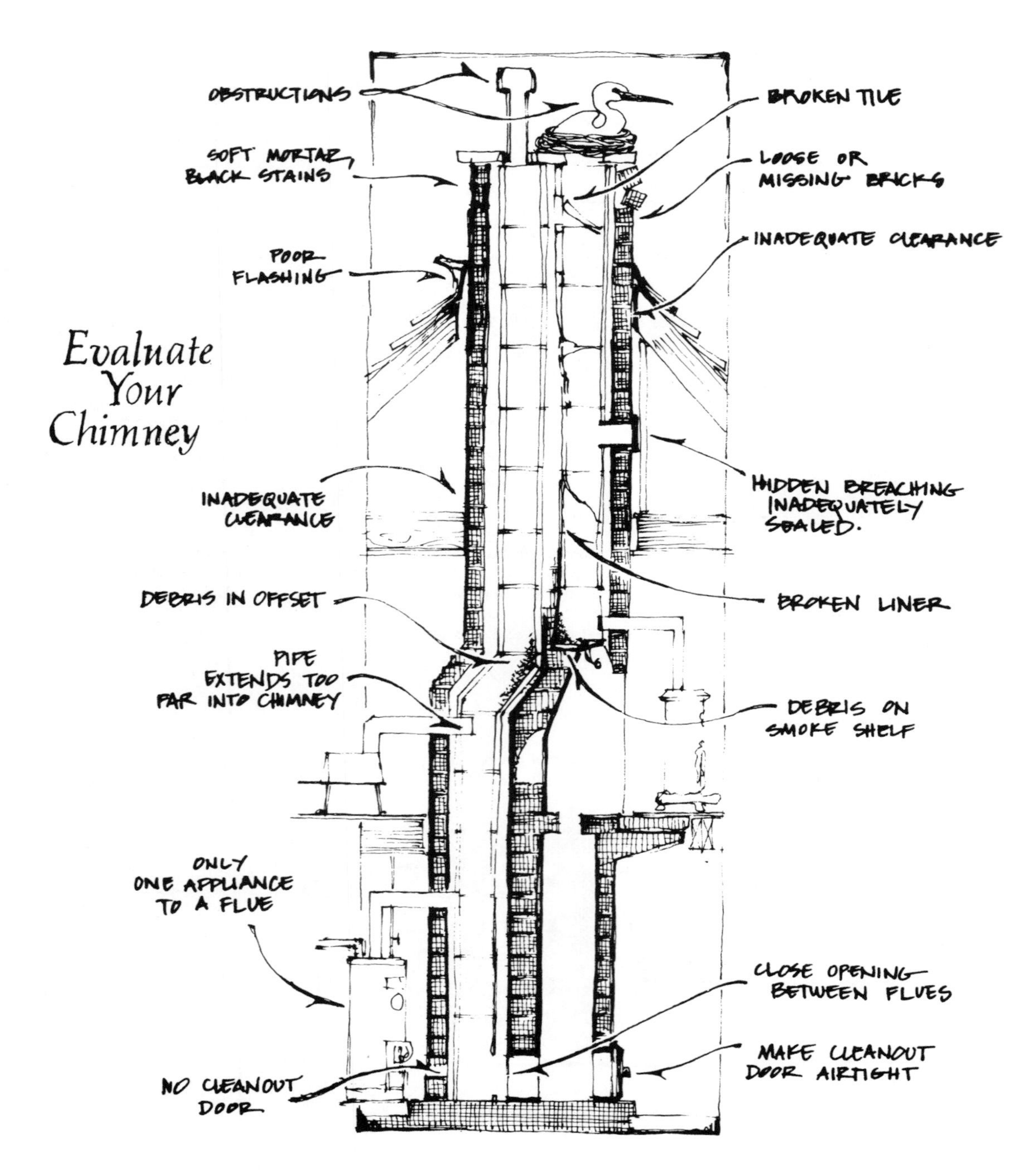

Evaluating an Existing Chimney

The plans for a new installation will include a look at whatever chimney possibilities are available. Many homes have an existing chimney in good repair that can be used. Examine it carefully to determine if it is suitable. If no bricks are missing and a sheet of paper is easily drawn against the thimble by the draft, most novice inspectors will give the chimney a passing grade. As you gain experience in what to look for though, many other criteria emerge.

What To Look For

Physical condition is the most obvious. The first thing to look for in any standing chimney is a clay liner. If you don't find a liner, don't plan to use the chimney without installing an acceptable liner of some sort. If a chimney has a tile liner, it should have no bricks or mortar missing. Deterioration in one spot probably means there are problems elsewhere, possibly in areas that are not accessible for visual scrutiny.

Step outside and observe the chimney height. Generally speaking, a chimney should be higher than any other part of the house. Years ago when wood was the primary fuel, chimneys were much higher above the roof than chimneys on recently built houses. Building codes today usually state that a chimney within ten feet of the peak of the house should be at least two feet above the peak. If the chimney is more than ten feet away from the peak, it should terminate at least three feet above the roof line and at least two feet higher than any part of the roof within ten horizontal feet. These minimum heights do not *guarantee* sufficient draft, and on some installations, greater chimney height is required for proper draft. The actual minimum height of a chimney has been the subject of some debate in recent years, but most agree that a flue should rise at least 12 to 14 feet above the stove's flue collar. Higher is always better, though, (at least if you're a chimney) and 20 feet will probably work better than 14.

Chimneys located on the interior of the house will work better and stay cleaner than exterior chimneys. Since the heat within the house helps to keep the entire column of the flue warm, flue gases also remain warmer. Gases that cool quickly, as may happen with an exterior chimney, will condense as creosote, and the velocity of the draft will diminish. Don't give up on exterior chimneys too quickly, though. They can be improved by building a protective enclosure around them, and special stove operation techniques that allow additional heat to escape up the flue can help, too.

The cross-sectional area of the flue should correspond to the flue collar size of the stove being used with it. The most common sizes of fairly new chimney construction are 8 x 12 inches or 8 x 8 inches. Generally, it is recommended that the cross-sectional area of the flue be no smaller than the area of the stove's flue collar. Some codes also specify that a flue size should not be more than 25% greater in area than the flue collar.

A flue size that is poorly matched to stove size will cause problems. If a large stove with fireplace doors that requires an eight-inch flue collar is installed on a 6 x 6-inch flue, there is a good chance that the stove will smoke when the doors are open. The inadequate draft may also cause the stove to act sluggish, make it difficult to get a fire established, and may result in a lower than normal heat output. These same problems may occur if

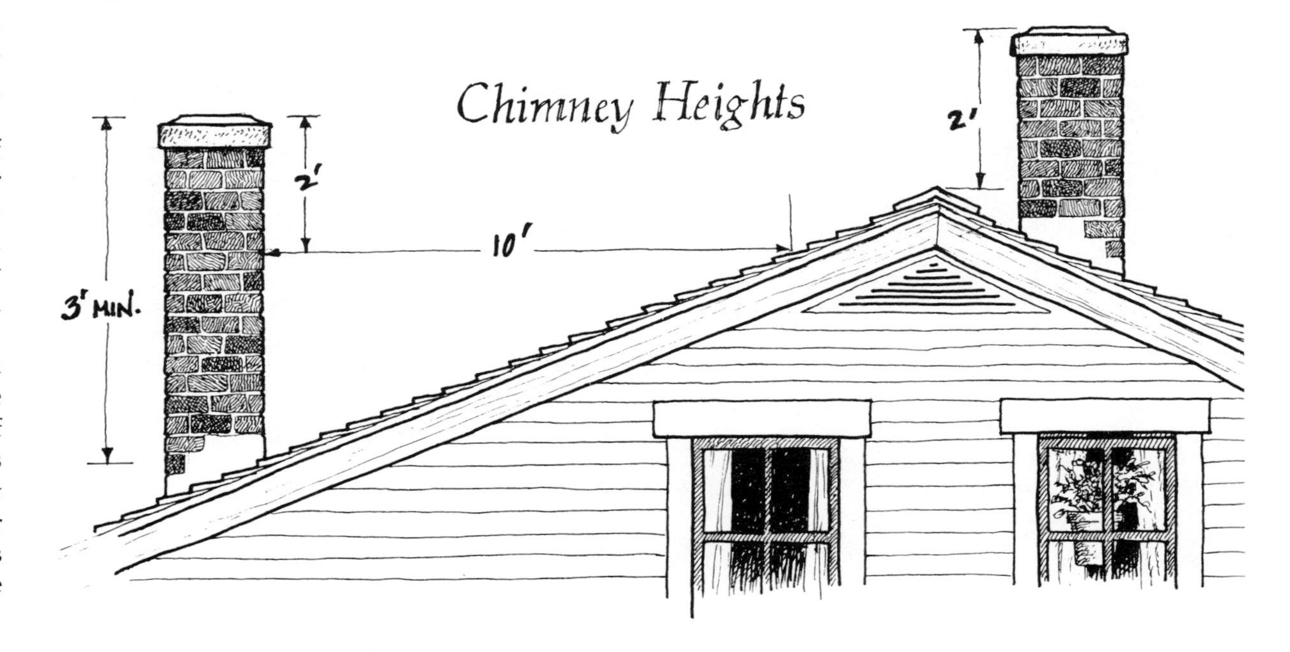

the flue is a proper size but a reduction, say, of eight to six inches is made in the pipe at the thimble area.

If a small stove with a six-inch flue collar is installed in an oversized flue with a 12- × 12-inch area or larger, problems will result as well. Small stove exhaust gases that enter a large flue will dilute and cool quickly, causing the same creosote formation and draft problems often witnessed in exterior chimneys. If the mismatched flue size happens to be in an exterior chimney, your problems are doubled.

Flues larger than 8 × 12 inches should be inspected particularly carefully before installing a wood or coal stove. Large flues of 12- × 12-inch size and larger are often old constructions and may have some structural problems.

Finally, a satisfactory chimney should be convenient to clean. Difficult cleaning access will result in procrastination of the cleaning job, and that is not safe heating. Typically, chimneys are cleaned from the top. New flexible cleaning rods allow cleaning access through the thimble and possibly even through the cleanout door. In any case, make sure the chimney can be cleaned easily, by a professional if necessary, before you commit yourself to using it.

Improvement Options

A careful evaluation that has found a chimney to be in need of repair presents several options. The most comprehensive, and probably expensive, route is to tear down the existing flue and put in a new masonry replacement. The replacement chimney may be either cement block or brick with a tile liner. The brick chimney is more attractive but will cost more. Some cement block chimneys are finished with a course of bricks around the section visible above the roof to give a more traditional appearance.

A well-built masonry chimney will be expensive, and the choice of location may be limited. At the same time there are several advantages. It may increase the value of your house; it will last indefinitely, and it will offer the peace of mind that comes with knowing that your heating system includes the safest flue type available.

In some instances, though, a complete replacement is not economically feasible. As an alternative, there are ways to line or re-line a chimney to improve both its safety and effectiveness.

One key element in relining is that the work be done carefully and well. Be sure to check with local authorities to find out whether there are code requirements to be met. The space between the existing chimney and the top of the new flue lining must be carefully sealed so no rain can get in that space, and no outside air can cool the flue liner. If there are bends in the existing flue, relining may be difficult.

When re-lining with stovepipe, a number of choices are available. A pipe with the inner surface coated to resist the corrosive effects that occur when coal and wood exhaust are combined with moisture should last a long time. Stainless steel is another good choice. Some manufacturers recommend what they call a "300 series" stainless steel, particularly if it is being used with a coal appliance. Standard stovepipe may also be used; it is less expensive but also may be less durable.

The Pre-fab/Coal Controversy

Some prefabricated chimneys may not be suitable for burning coal. A number of stoveowners have found this out when they have switched from burning wood for several years to coal and continued to use the same flue. The exhaust from coal smoke has a high sulfur content—above 1%—compared to wood. When the sulfur particles mix with moisture in the air as they rise up the flue, highly corrosive sulphuric acid is formed. When left on the interior walls of the chimney for prolonged periods of time, the acid will begin to destroy the metal. The corrosive action can be retarded by frequent brushings of the flue, and burning a good quality anthracite with a low sulphur content will help too.

Prefabricated chimney manufacturers are in the process of developing better grades of stainless steel liners for use with coal-burning appliances. Recent manufacturers' recommendations have been for "Type 300" stainless steel when burning coal but the recommendations may change. Before buying a prefabricated chimney for use with a coal stove, check with the manufacturer. Also check with your local building inspector to determine if his regulations specify certain brands of chimneys for use with coal.

One stoveowner reported that a carefully done re-lining, using standard smokepipe, was in good condition after six years. Others have told of deterioration within one year.

Prefabricated, double-wall insulated pipe is also used. In fact, it is the only material allowed for re-lining in some areas. It is expensive, and the existing flue would have to be quite large to accept it, but it should work well.

An option, used for years in Europe but relatively new in America, is to re-line with a masonry product. A deflated tube is put in the existing flue, and the tube is filled with air. The lining material is poured into the space between the existing chimney and the inflated tube. The new liner is solid and smooth. The lining material is supposed to strengthen the chimney and insulate the flue.

A popular alternative to the masonry chimney is the manufactured, or prefabricated, type. Two main types of manufactured metal chimneys are available. The first is referred to as "double-wall insulated," and basically features a steel pipe within a steel pipe, with the one-inch space between the two components filled with a mineral insulation.

The other type of manufactured pipe is generally called "triple-wall," and features a pipe-within-a-pipe-within-a-pipe arrangement. An air space is left between each pipe to act as the insulating barrier.

Prefabricated chimney manufacturers offer a full line of prefabricated accessories that allow considerable flexibility in how the chimney is installed and simplify the installation. Manufactured chimneys are not inexpensive, but will probably be cheaper than masonry construction. The main cost advantage of the manufactured chimney is the labor charge for installation: You may be able to do the installation yourself, and if not, a competent installer can do the job easily in a day.

Chimney Caps

"Improve draft, eliminate creosote build-up, promote higher efficiency, prevent down-drafting, extend the life of your chimney. . . ." These phrases are often used to sell chimney caps. Do these things really happen when you install one?

Good draft, a minimum of creosote build-up, and efficient operation are usually related to having a well-constructed chimney, a good fuel supply, and good stove-operating techniques. If a particular draft problem persists, a chimney cap may help. However, a conservative suggestion would be not to use a cap unless there is a good assurance that it will work. Local experience is a good guide when selecting a cap to solve a particular problem. Talk to people who have used the product.

For solving wind-related problems, a crown-shaped cap has proven effective. These caps can prevent down-drafting and stabilize draft under changing conditions.

A cone-shaped cap also can be effective in preventing down-drafting. Another use for this cap is to reduce the opening at the top of a large flue to about the size of the flue collar. This may improve draft and has resolved some problems where marginal draft was indicated.

Prefabricated Pipe

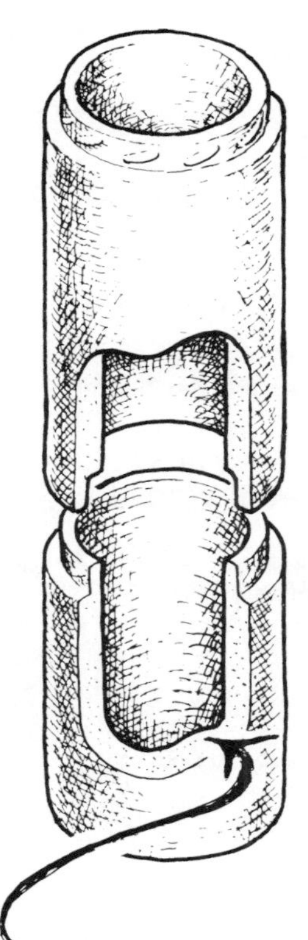

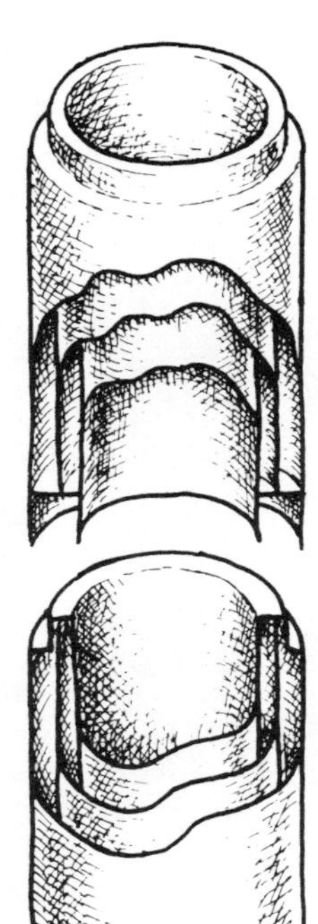

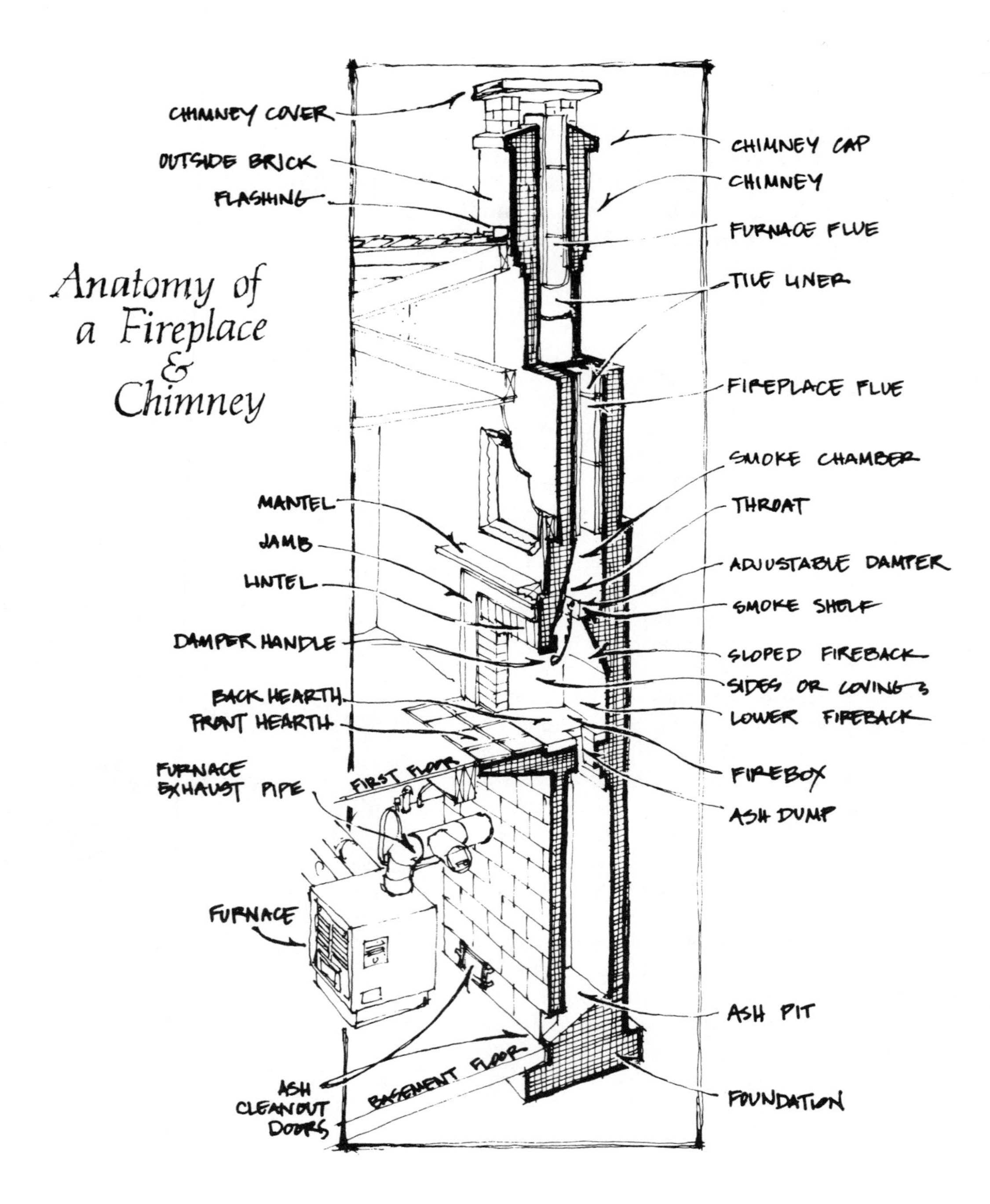

Anatomy of a Fireplace & Chimney

Many people use flat masonry caps, supported at the four corners about 8 to 12 inches above the top of the flue. It keeps out most of the weather, is durable and attractive, and is a good option.

There are some reservations about caps that spin in the wind or change direction with the wind. Snow, ice, or creosote may interfere with their operation. It is possible that a turbine-type cap could make shutting off air to the flue in case of chimney fire more difficult.

Many chimneys work well with no cap. Install one only if you have a draft situation that cannot be corrected through a change in fuel or operating technique.

The Fireplace Option

Some homes will have an unused fireplace as well as an existing chimney. It makes sense to put it to use. Until the nineteenth century a fireplace was the sole source of warmth and cooking in most homes. Then, iron stoves became recognized as technological improvements. For the next hundred years or more the fireplace fell out of popular use with the exception of a ceremonial fire a few times a year.

Then came the energy crisis of the 1970's. With the cost of home heating suddenly raging beyond the confines of the family budget, families across the country began scrambling to caulk windows, add insulation, and explore alternative heating methods. Suddenly, the almost abandoned fireplace in the family room took on a new significance. There, already in place, was a way to generate heat. Once again the fireplace became a center of activity. Somehow, though, the romantic recollection of fireplace charm was not quite right. You had to sit close to it to stay warm, and it seemed to use a lot of fuel. Wood, which we spent hours gathering, sawing, splitting, and hauling, was consumed at an alarming rate.

Even the banked fire was cold long before morning. We invested in tubular grates, blowers, glass doors, and machines to make logs out of newspaper in an attempt to improve the efficiency . . . but the fire was still out long before morning. Slowly, we began to read and hear and finally understand the shameful truth, hidden by centuries of romance: The fireplace is an energy hog.

Laboratory studies, backed by our own observations, produced statistics that showed that fireplaces contribute negatively to a heating plan by drawing off warm air from the rest of the house all through the night, long after the last embers had died. That was the same effect as leaving a window open all night. The gadgets we had so hopefully purchased raised the efficiency only slightly. Statistics consistently showed a fireplace was at best only 10% efficient, with the remaining 90% of that hard-earned heat lost up the chimney. The tubular grates and glass doors raised the 10% figure only a few points.

At the same time, other ways to keep warm efficiently were evolving. Foremost of the alternatives was the airtight stove which would burn wood, coal, or both. With efficiency levels of 50% or higher, airtight stoves offered more heat from the fuel source and made it last longer. It wasn't long before a marriage between the airtight stove and the fireplace was imminent.

Installation Methods

There are several ways to connect your stove to a fireplace. Generally, it is best to install the woodstove on the hearth in front of the fireplace rather than recessing the entire unit into the fireplace cavity. A radiant stove heats your home by allowing air currents passing by it to carry the heat away. If five of the six stove surface plates are tucked back in a fireplace, the efficiency potential is significantly diminished.

An easy and fairly common method of venting the stove into the fireplace flue that was once popular is no longer recommended. This method involved covering the entire fireplace opening with a metal plate or mineral board and establishing an airtight seal around the perimeter. A hole corresponding to the size of the stovepipe was cut in the metal plate, and the pipe simply passed through the plate from the stove and ended. This method sometimes resulted in poor stove performance; the large fireplace cavity could quickly cool the flue gas and result in excessive condensation as creosote and sluggish draft. Some manufacturers have even reported that the poor draft could result in small backpuffing "explosions," some with enough force to push the plate, as well as the stove, slightly forward.

A preferred method involves replacing the existing fireplace damper with a metal plate. The perimeter is sealed to make it airtight and a hole is

Make It Easy on Yourself

In terms of sheer frustration, few experiences rival that of cutting stovepipe. The instructions which accompany your newly-purchased pipe (if indeed there are any) will be deceptively simple. "Snap together the sections and cut to desired length using tin snips." Several hours later, your floor littered with pieces of mangled pipe, your hands a mass of pulpy flesh, and your wife having taken the children to the safety of a neighbor's house, you will find yourself with arm poised to hurl the tin snips through the picture window. It is rumored that most of the swear words in the language were invented by people cutting stovepipe.

At Vermont Castings we formerly limited our advice on cutting stovepipe to post mortem marriage counseling and instructions on how to replace picture windows. Then one day we discovered a tool which enabled us to cut stovepipe in a reasonably straight line without leaving finger flesh on the jagged edges. This tool, that appears to have been invented by someone who in a pique of anger put the blades of his tin snips into a vise and then whacked them with an eight-pound sledge, has done for the hands of America what the first safety razor did for the face. To whomever invented this minor masterpiece, the stoveowners of America salute you.

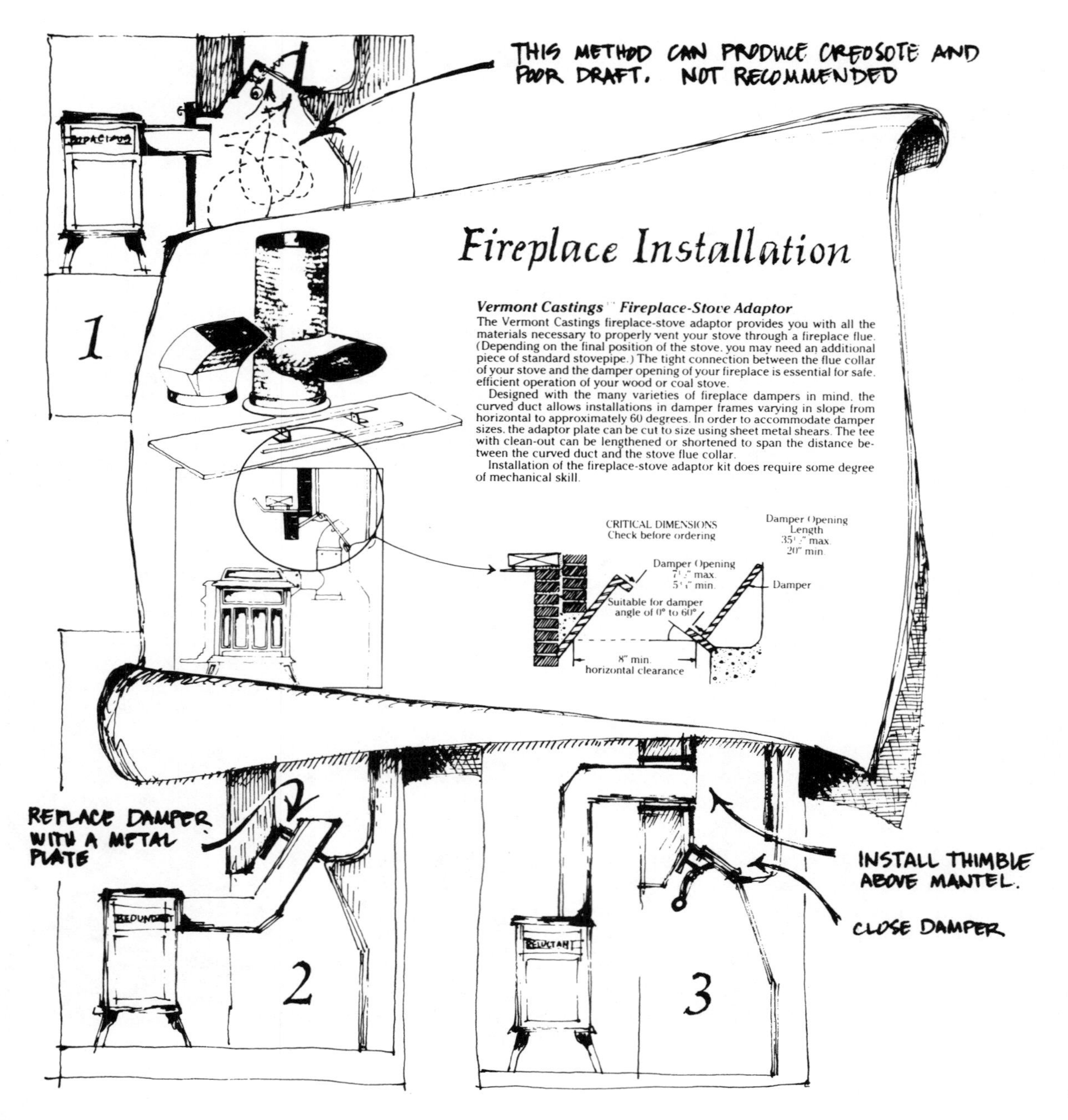

cut in the middle to accept the pipe. The pipe extends through the metal plate and into the flue for a distance of 6 to 12 inches.

This installation is not technically complex, but does present a challenge because of the cramped work area and the accumulated soot that will hinder your progress. You can save several hours of dirty work and several levels of your patience inventory by tackling the installation with the help of a specially made fireplace adapter kit.

A third installation option is to bypass the fireplace opening altogether. This means installing a thimble above the mantel in the chimney itself. The stovepipe is then vented vertically from the stove and angled into the thimble. The damper should be permanently closed, or the fireplace opening itself may be bricked in. Don't forget to make some provision for a cleanout door if you choose the latter option. Any part of the mantel within 18 inches of the stovepipe should be protected with a heatshield.

The actual installation of the thimble may require that you seek professional help from a mason. The existing brick and mortar must be removed by a cautious drilling and chiseling procedure, and great care must be taken to avoid damaging the internal chimney lining. This job is best done by someone who has done it before and knows exactly what needs to be done.

Extend the Hearth

Once the stove is placed on the hearth and the pipe has been properly connected to the flue, stand back and admire your efforts. At the same time, examine the floor area around the hearth to determine if it needs additional protection. A combustible material, whether it be floor or carpet, must have a suitable protective covering where it is close to a stove. Remember, the standard area recognized as requiring this protection is within 12 inches of any stove side, with the exception of sides that include loading doors. These sides that can open need 18 inches in front of them.

Use the same procedure described for hearthpad construction in Chapter Two. If your existing hearth is flush with the floor, you may wish to cover the entire area, existing hearth and floor, with the portable hearthpad described previously.

Stovepipe for Everyone

Your chimney has been evaluated and has passed the test, or you have repaired it. The stove has been wrestled into place on the hearth, leaving you with a strained back and a neighbor who will forever after be suspicious when you offer "a couple of beers in exchange for a little work." All you need to do now is make the connection, stove to chimney. This part of the installation should not be treated lightly, and we should spend some time talking about the connection, or stovepipe.

Stovepipe is the critical connection between your stove and the chimney; it is usually the weakest part of an installation and requires careful attention.

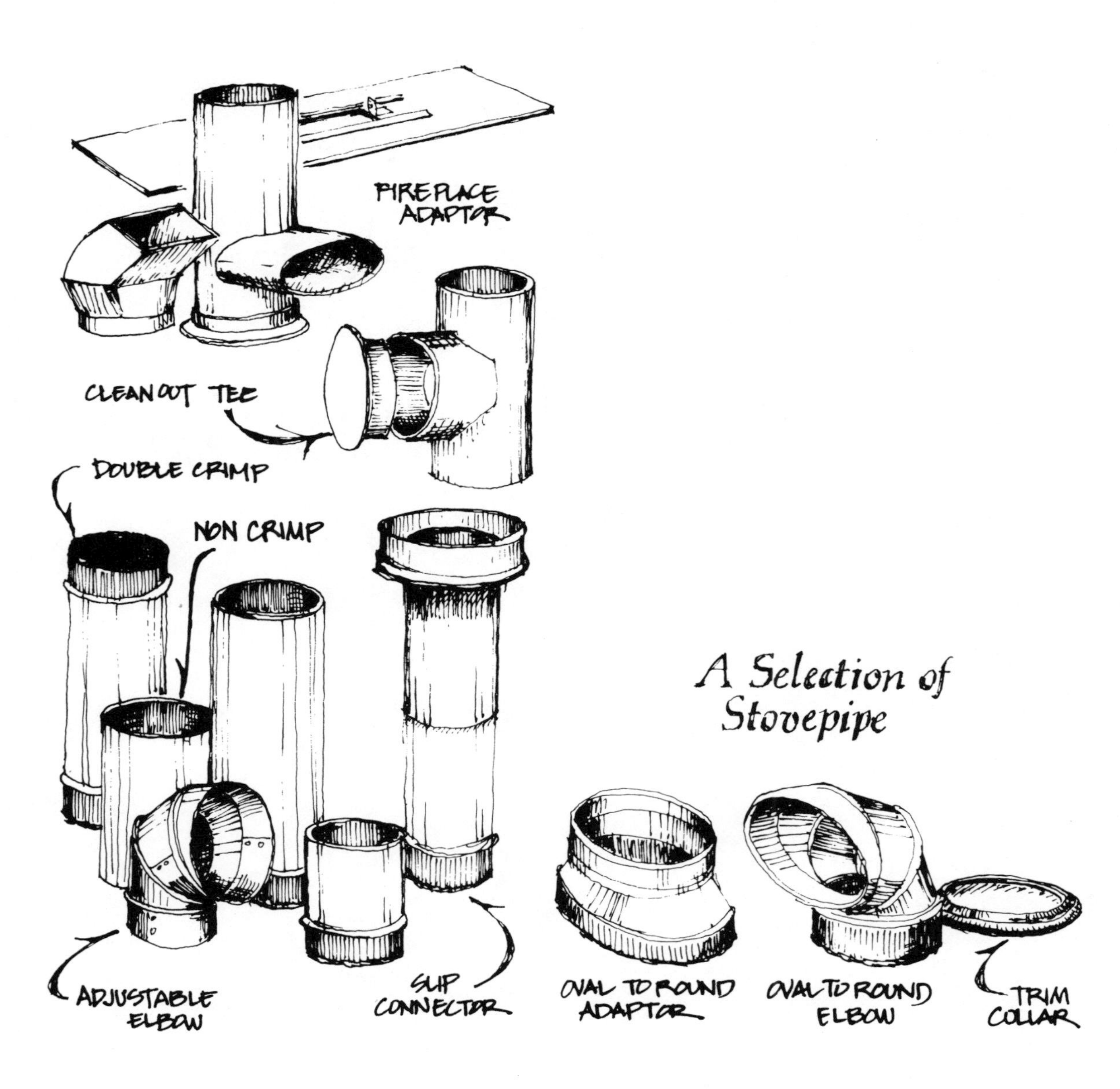

A Selection of Stovepipe

Which End Is Up?

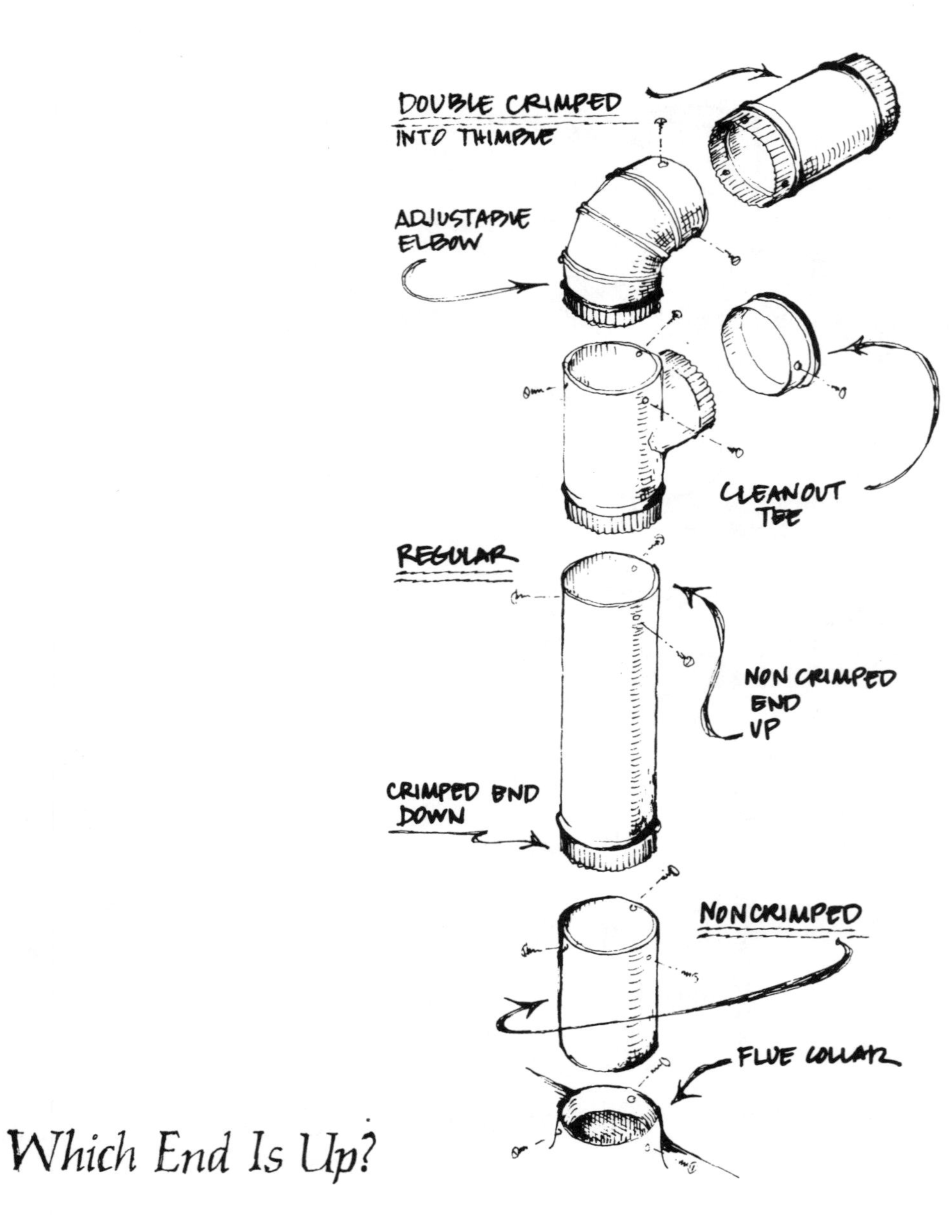

Stovepipe may be made from sheet steel, stainless steel, or galvanized steel. Galvanizing is the process of coating sheet steel with a solution, usually zinc-based, to prolong its life. Most installations use standard sheet steel pipe.

The thickness of stovepipe is expressed by the term "gauge," and the lower the gauge number the thicker the pipe.

The gauge of the stovepipe will determine how long it will last and how well it would fare under the extreme condition of a chimney fire or constant creosote accumulation. Traditionally, hardware stores have sold 28-gauge stovepipe. As woodburning became more popular, however, 24-gauge became more widely available. Now, 22-gauge pipe is starting to be required in some areas. The heavier the gauge of the pipe, the longer it will last.

Pipe also comes in a variety of colors, primarily silver, shiny black or blue-black, or flat black. The choice of color is an individual matter of aesthetic preference although dark pipe will radiate a slightly higher level of heat.

Stovepipe manufacturers "crimp" one end of a section to reduce the circumference to a point where it will fit inside an uncrimped section of pipe. For installation flexibility, some pipe sections have no crimp at all, and others have a crimp on either end. Stovepipe elbows, both adjustable and rigid, allow you to make corners and a variety of angles. There are also specialty pieces, such as "boots" or starter pieces that attach directly to the stove.

Pipe comes in a variety of diameter sizes, with the most common being six-, seven-, and eight-inch. "Six-inch pipe" is pipe with a diameter of six inches, not a piece that is six inches long.

How Much Will You Need?

Plan your stovepipe shopping list with the attitude that more is always better than not enough. Return any leftover pieces in original condition to your supplier for a refund or credit, but don't let yourself run out in the middle of the installation.

Excluding the fireplace installation that has been discussed separately, there are three standard configurations of pipe in an installation. The first, when the wall thimble is directly in back of the stove, is the easiest to manage. You will need only one section of pipe, with the crimp on the thimble side. The end of the pipe entering the stove may or may not need a crimp; this depends on your stove. Secure the pipe to the stove's flue collar with sheet metal screws.

The second pipe configuration is called for when the thimble exists in the wall above the stove. This type of installation will require a vertical rise of pipe, an elbow, and a horizontal run into the thimble. If the thimble is offset, more than one elbow may be needed. It is always best to keep the number of elbows to a minimum, however. Horizontal runs of pipe should be kept to a minimum also. Draft problems may arise if the run exceeds four feet.

The third installation possibility exists when the stove is to be vented into an installed prefabricated chimney with the thimble directly above the stove. Two special pipe pieces will simplify life for this type of installation. The first is referred to as "slip-connector," and is a telescoping section of pipe that extends from approximately two to three feet. Pipe installed with the slip connector is easily installed initially, and is not difficult to take down for cleaning access.

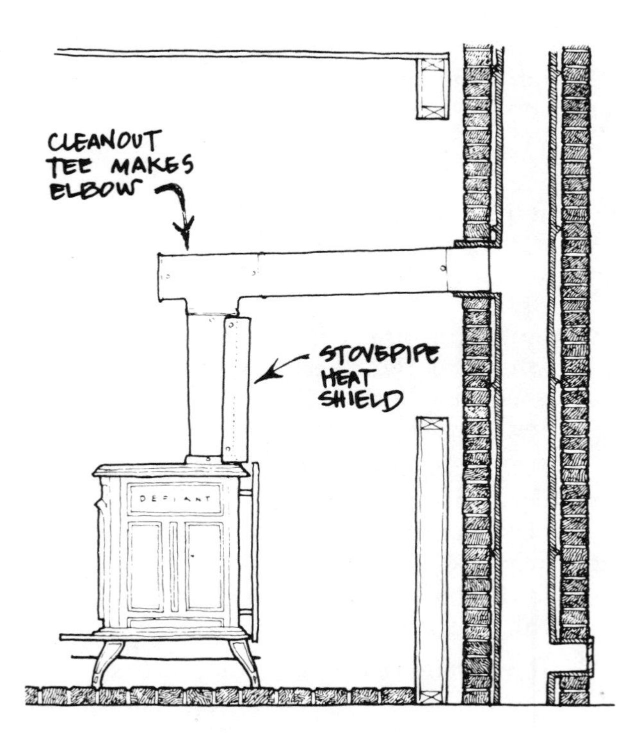

Thimble Above Stove

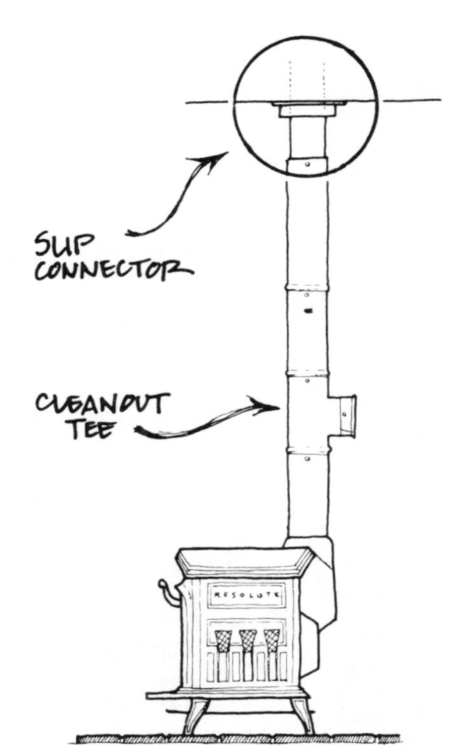

Prefabricated Chimney Above Stove

The second special piece of pipe that may be used is the "cleanout tee." This piece is installed below the slip connector and has a removable plate that allows regular inspection of flue condition. Cleaning can also be accomplished through the tee with a round chimney brush and flexible rods.

Working with Stovepipe

Handling stovepipe requires some special tools and a great deal of patience. Assemble the following supplies before attempting a stovepipe session:

- Gloves to protect your hands
- A cutting implement—lever snips are preferred, but sheet metal snips will do
- Felt-tip pen for marking
- Flexible rule, like tape measure, for measuring
- An electric drill, two 1/8 drill bits, (one in case you break the other) and a couple of dozen sheet metal screws (#10 x 1/2 inch) for securing the pipe sections together

With these tools and with the patience mentioned above, you should be able to complete your installation. As with any do-it-yourself project, working slowly and measuring carefully are the cardinal rules. Here are some additional hints to help you through the process as well:

- Interlocking stovepipe really *will* snap together; using six-inch pipe is harder than eight-inch, but it can be done. Start at one end by engaging the lock, hold it in place with your first two hands, and work your way down the pipe with your other two, fitting the lock together as you go.
- The crimped ends of all pipe should point toward the stove. Any liquid condensation will then end up back in the stove.
- Fasten all pipe sections together with the sheet metal screws. Secure the pipe to the flue collar as well.
- If your stovepipe enters a wall thimble, secure the fit of the pipe with the use of two eye-screws, a length of wire, and a spring.
- If an increaser is to be used, it should be located as near the thimble as possible.
- Vertical lengths of stovepipe should not be excessive; there are varying opinions on how much is too much, but less than 12 to 14 feet is the amount acknowledged to be acceptable. (In the case of a cathedral ceiling, it would be preferable to have the metal chimney extend a greater distance into the room.)
- Stovepipe dampers were common in the days before stoves became airtight and were employed to help control the rate of burn. They are seldom necessary with airtight stoves that control rate of burn at the air inlet.
- Stovepipe heat exchangers do not work well with modern airtight stoves. Heat loss up the flue is minimal anyway, and the exchangers tend to encourage creosote.

Stovepipe may last four years or may become useless after a single season; the longevity depends on how the stove is operated and how regularly the pipe is cleaned. Inspect it periodically, and replace it as soon as there is any suspicion of deterioration.

AN EXTRA PAIR OF HANDS IS OFTEN HELPFUL

Enamelled Stovepipe

If you plan to purchase an enamelled stove, you may wish to install enamelled pipe for a more aesthetically appealing look than standard pipe will provide. Usually it is best to stay with pipe manufactured by the same company that builds the stove so that the colors will match.

Enamelled stovepipe is the most heat resistant, durable, and expensive type of single-wall pipe available. The tough enamel coating is usually applied on both the interior and exterior of standard diameter stovepipe. It has a high AR (acid resistance) value, and will resist staining and corrosion caused by creosote.

The exterior surface of enamelled stovepipe can be cleaned with a mild non-abrasive cleansing agent. This should be done when the stove is cool, and all traces of the cleaner should be removed before refiring the stove. The interior of enamelled stovepipe should be inspected and cleaned regularly in the same fashion as regular pipe.

The largest drawback to using this pipe, other than the expense, is its lack of flexibility. Because of the sheer weight of the enamel, it is extremely rigid. While this makes for a durable pipe, it also causes difficulty in cutting to desired lengths. Most manufacturers have overcome this problem by the use of a telescoping section that allows a variance of a few inches in length.

Precise measurements are important in ordering the correct length of enamelled stovepipe. Avoid using this type of pipe in installations where there are odd angles, such as a connection to a thimble that is off-center in relation to the stove. When the pipe is to be installed so that it exits off the rear of the stove into a fireplace or wall thimble and not be highly visible, you may want to use standard non-enamelled pipe. Enamelled elbows are non-adjustable although some manufacturers produce elbows of more than one angle, i.e., 45° or 90°.

All sections of enamelled pipe should be connected with sheet metal screws as with standard stovepipe; this may require you to drill holes in the pipe. A high quality carbide bit should be used for drilling. The pipe should be assembled before it is drilled to make sure it has been measured correctly. Be very careful when handling the pipe, since it will chip easily if struck with a hard object.

Generally, the clearances for enamelled stovepipe are the same as those for standard stovepipe. However, because of the difficulty in drilling, it may not be recommended by the manufacturer to install stovepipe heatshields when necessary to reduce clearance to a combustible wall. It is always best to check with the manufacturer for his recommended clearances.

Summertime Also Is . . .

Vermonters like their stoves, and put a lot of energy into the entire heating activity. But that's not all they do. A favorite summer evening activity is porch-sitting, the same pastime enjoyed in hundreds of small towns across the country.

Porch-sitting in Vermont is a little different than in some places though. Many parts of the state are still rural, and life on a remote dirt road can be quiet. An entire evening may pass with only a car or two passing by. Since a part of porch-sitting involves waving to people you know and making comments about them that they can't hear, it doesn't take many quiet evenings before you realize you are missing part of the fun. The answer is to get in the car and go for a drive, usually past your neighbor's porches. This mobile version of porch-sitting allows you to experience the tacit social interaction of the pastime even though you are not on your own porch.

There seem to be one or two nights each summer when the entire town chooses to go for a drive at the same time. This means that you will find few porches occupied but more traffic than normal on the roads.

In addition to porch-sitting, every small town has its favorite special events.

Summertime in Randolph, Vermont is often described as "a two-week period that occurs between the Fourth of July and Labor Day." The season is one of great activity as everyone seeks to rid himself completely of the last vestiges of cabin fever. Gardens are in by Memorial Day, although the threat of a killing frost is a reality throughout June. More than once, one will have to cover the tomatoes at night if he has any hope of spaghetti sauce for next winter.

By the Fourth, summer is in full swing. The day is celebrated in a suitably Small Town American way with a footrace, potluck dinners, the world's greatest parade, the Hoxie Brothers' circus, a mandatory visit to your favorite (and secret) swimming hole, little league games, and finally fireworks. July drifts into August where the big event is the Vermont Castings' Owners's Outing, a unique celebration of its customers by an appreciative company. The town of Randolph swells to up to three times its normal size as devoted owners stream in from around the country to see where their stoves were born.

Labor Day provides the second bookend to summer. At this time, the mighty zucchini is justly feted. The prolific member of the squash family thrives in the short Vermont growing season. We overcompensate by planting ten times more than we can conceivably use. After unloading armloads on our friends and feeding the pigs (we've even established a "free zucchini" box in the Vermont Castings' showroom for out of town visitors) we still find ourselves with enough zucchinis to justify a harvest festival. Zuke-carving, bobbing for zukes, zuke launches and other traditional summer squash activities are held within a culinary context that runs the gamut from zucchini soup to zucchini ice cream.

Two weeks later comes the renowned Tunbridge World's Fair where Central Vermont's largest zucchini earns a blue ribbon amidst a backdrop of flimflam and hootchie-kootchie. But by now the first killing frost has spread its pall, and summer is officially over. Looking back on a montage of church suppers, fishing trips, fiddlers' contests, country auctions, and craft fairs it seems indeed to be an event-packed "two weeks."

Above all, summertime is a time of indulgence. In the stove business we prepare for the long-awaited fall rush by regaling the rookies at the company with sagas of some of the perils and pitfalls out in Stoveland. Here are some of our own indulgences, a collection of the more absurd encounters we have ever experienced.

A Case of Mistaken Identity

A particularly busy Monday morning in October had forced a trainee onto the phones to deal with customers. One of his first calls was a request for a "18-gauge, over-sized steel gasket." After checking with a supervisor, the fledgling Customer Relations representative dutifully informed the customer that our gaskets were made of either glass fiber or asbestos. An awkward silence, then

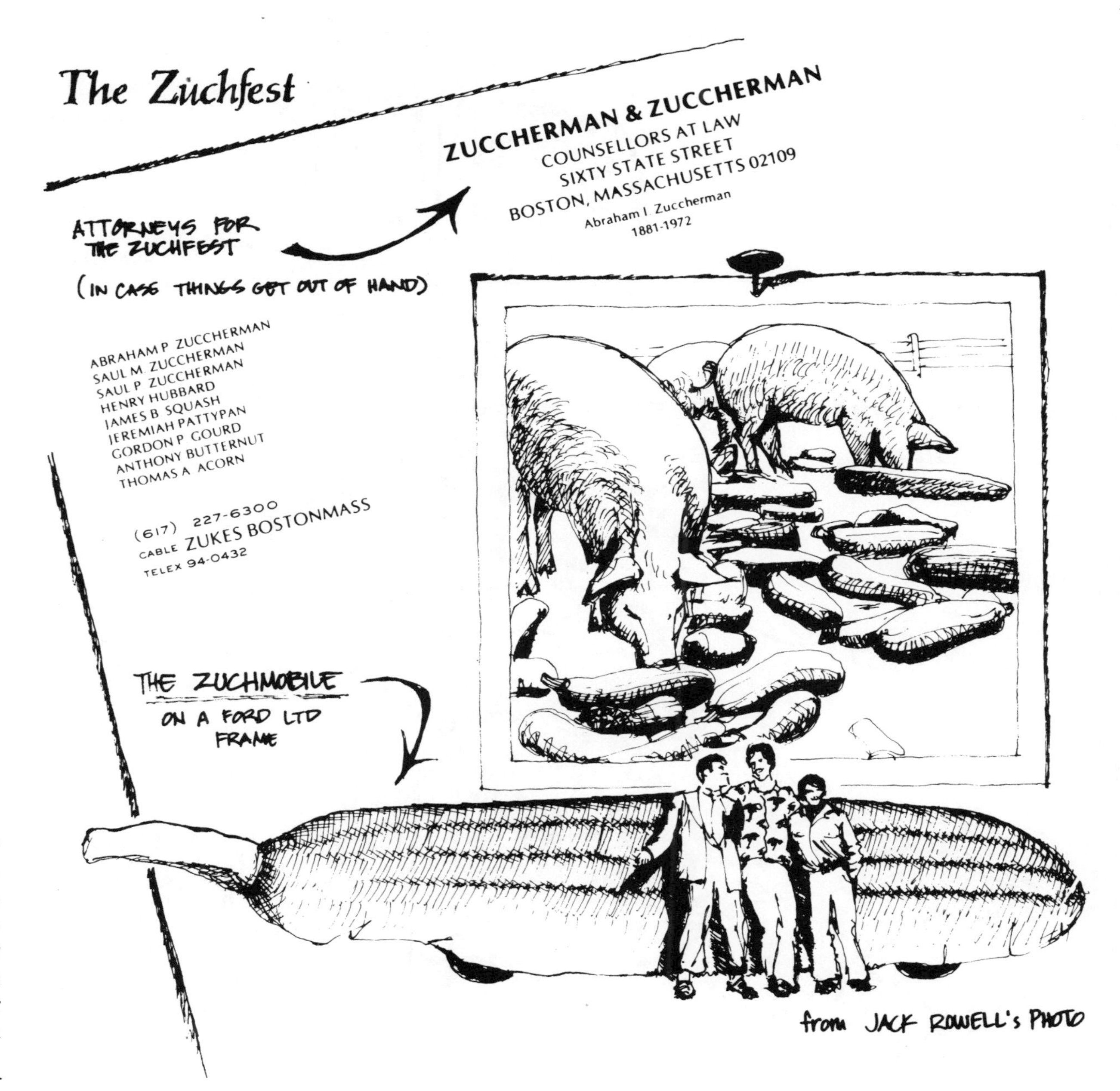

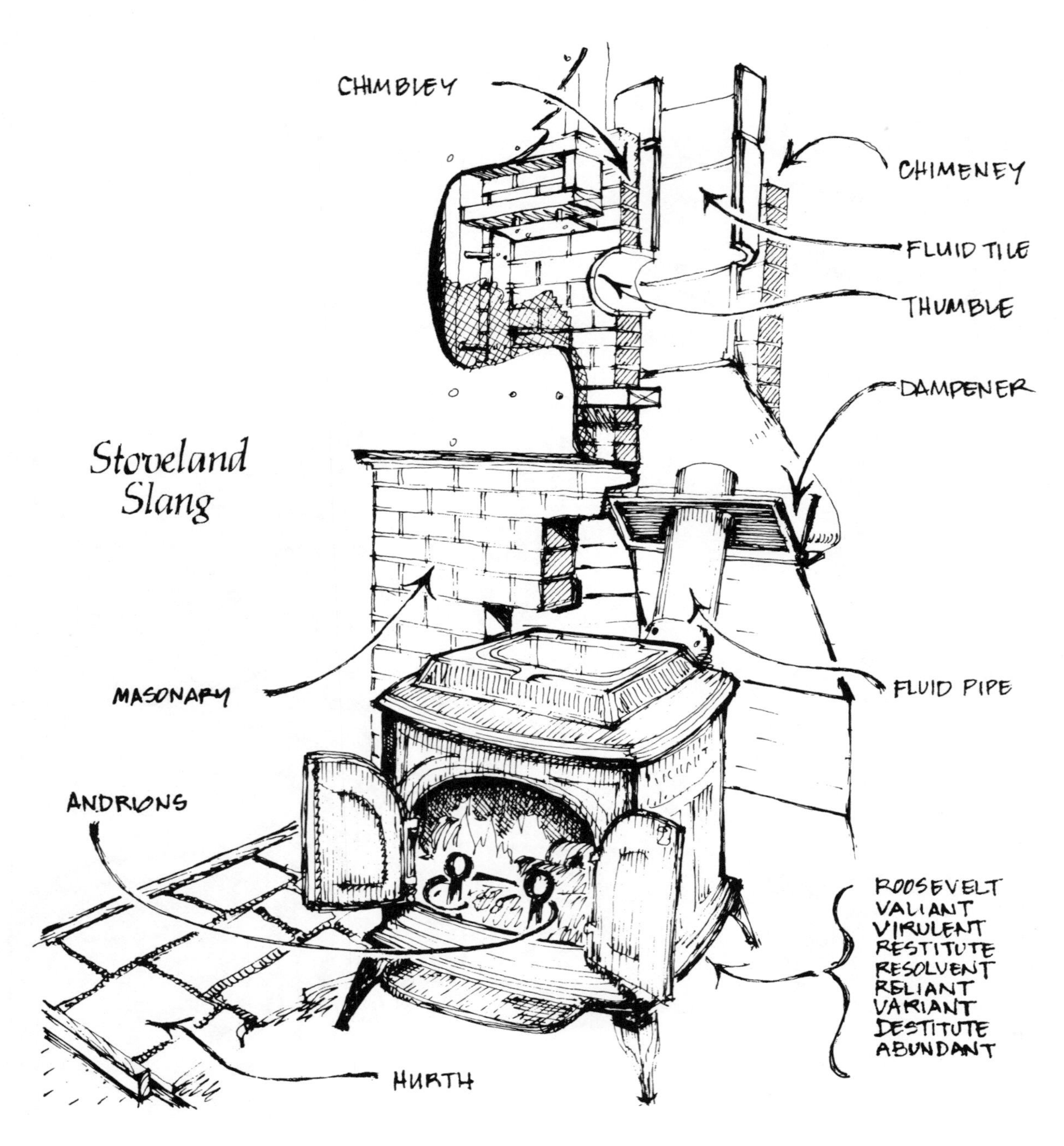

the customer requested an "18-gauge over-sized steel gasket." Again the Vermont Castings' representative patiently explained the available gasket options. Seeking additional information, he then asked which model the gasket was for. The customer responded that it was for a very large woman. After a silence that communicated total confusion, the customer elaborated, "Yeah, she must weigh at least 350 lbs." (This happens to be about the weight of a Defiant Parlor Stove, a point which further confused the issue.) The call was finally transferred to a supervisor who determined that the bereaved party on the other end of the phone had thought he was calling the Vermont Casket Co.

* * *

At a wood heat convention in New Orleans several members of the Vermont Castings' staff found themselves enjoying the best hotel service any of them had ever experienced. Not only was the hotel staff courteous and attentive, but they performed even the most mundane tasks with an exuberance that bordered on the theatrical. Finally one of our staff members could contain himself no longer and blurted out to a busboy who had just completed a thankless task amid a flourish of smile and song, "I can't believe how enthusiastic the staff of this hotel is! Everyone we meet is singing, dancing, smiling, and cracking jokes while he works!" The busboy smiled from ear to ear and answered sheepishly, "Tell you the truth, sir, when we heard you were Vermont Castings, we thought if we played our cards right, maybe one of you would give us a part in the movie y'all are making."

Summer Checklist

The living is easy, right? Instead of hauling ashes, you can mow the grass. Still, this is the season for packing away the stove and forgetting the task of staving off winter's chill.

The new stoveowner will be active the summer before his first heating season building coal bins and wood boxes, perhaps even undertaking major renovations to put in suitable chimneys and hearths. Keep in mind that the care that goes into these initial efforts will go a long way towards determining the ultimate enjoyment one receives from this pastime. The pleasure of a crackling fire is great, but it pales in comparison to the feeling of security which comes from a solid, well-designed installation.

- ☐ ***Put ashes on the garden.*** *You have been very patient about this. Remember, don't do it until just before tilling.*
- ☐ ***Scavenge kindling.*** *This is the season when all the do-it-yourselfers of the world are creating mountains of perfect kindling which will find its way to the dump unless you get it first.*
- ☐ ***Relax.*** *By now next winter's wood supply should be cut, split, stacked, and covered. Now it is your turn to sit back with a cold drink and watch the sun do its work. Gloat a little. You've earned it.*
- ☐ ***Deodorize your installation.*** *Sudden climatic changes will create reverse drafts which will bring Mother Nature pouring down your chimney and out into your living room. Unfortunately, if Mother Nature wafts over vast creosote deposits she will smell much worse than if she travels down a clean flue.*
- ☐ ***Coalburners, clean your chimney, again!*** *Coal soot and fly ash contain minerals that become strongly acidic when mixed with water. Heavy summer humidity combined with an uncleaned flue can be a lethal combination, particularly with any kind of pre-fab chimney. Consider this fair warning.*
- ☐ ***Check masonry and roof flashing.*** *You don't want to do any work of this type, but if you must then you might as well do it during warm weather.*
- ☐ ***Reserve your local sweep.*** *From September 1 onward your local chimney sweep will be in demand. If climbing onto the roof to clean flues is not your idea of a fun Saturday, then call your sweep, now.*
- ☐ ***Put your stove to bed.*** *First of all, there are probably a few major repairs that slid by during the heating season. Fix them now, particularly if replacement parts are required. Nothing is quite as frustrating as trying to get a new do-hickey from a stove manufacturer in October. Do it now. You will thank yourself in the fall.*
- ☐ ***Pamper your stove.*** *Check the seams, seals, and door latches. Apply some WD-40 or a similar lubrication to all moveable parts. Remember, (during the summer) your stove's biggest enemy is rust.*
- ☐ ***Break out the kitty litter.*** *Clean out ashes throughly, using a flexible vacuum cleaner hose to get behind hard-to-reach baffles. Then sprinkle a liberal portion of kitty litter in the base of your stove to absorb moisture and odor. Be sure to keep all doors and griddles closed or your friendly household cat may defeat this strategy.*
- ☐ ***Oil your griddle.*** *Any unpainted, exposed surfaces, such as a cooking griddle, should be protected from moisture by a thin coat of cooking oil. Do yourself a favor, and don't set a plant on your stove. You will be guaranteed a rusty ring in fall.*

3
Fall

School Days
A Buyer's Guide
A Coal Primer
Welcome to Stove School
The Stovetop Chef
Fall Checklist

The swallows and robins are smart enough to heed the warning of the falling leaves. The days can be inspirational, but the message is blared from every fiery maple: Don't be fooled by the warmth of the daytime, the frosty evenings foretell what lies ahead.

School Days

On Memorial Day it seems as if the coming summer will last forever. The project list is a mile long and the time to complete it infinite. By the Fourth of July the season has blossomed to full flower. However, before the smell of fireworks has faded from memory, summer is over. The signal may be the first chilly evening or a flash of red in a tree, but the slide has begun. Labor Day slips past, school begins, and the heating season advances relentlessly.

September brings many tourists to Vermont to view the spectacular foliage. Increasingly in recent years, a new group of visitors has been brought to our fair state as the Freshman Class comes to Randolph to attend Stove School.

Stove School is not an advanced institution of higher learning so much as a simple explanation of the principles of stove operation. We found that too often people would make the colorful pilgrimage to Vermont, pick up their new stove, then be calling back the next week with the most outrageous questions imaginable. A staff joke concerned the person who picked up his stove on Saturday, installed it on Sunday, then in complete disregard of prudent break-in procedures, immediately built a roaring blaze kindled by charcoal lighter fluid and his unread operation manual.

What was not funny about this sterotype was the resulting cracked bottom that we would hear about on Monday morning. By the time the customer made a second trip to Randolph to have his stove repaired, the novelty of the foliage had become secondary to the needless effort of re-installing a repaired stove.

Before long we realized that it was not the new stoveowner who was negligent so much as ourselves for not explaining the basics of stove operation when we had a chance. Thus, Stove School became mandatory.

The ability to burn wood or coal is not an innate talent of the human species. Many a grizzled oldtimer has attempted to avoid our modest classroom by claiming vast experience with the trusty Franklin or potbelly, but we now know better than to let him get away. Wives and children are given the course as well. The ten-minute investment in the classroom can pay dividends for many years.

When we firmly but politely suggest that the new stoveowner go through Stove School, we are hoping to bring expectations to a realistic level. Even a thermostatically-controlled stove will not be able to keep a room at an even 72°. The stove that *can* burn overnight will not do so if it has only two logs in it. The novice coal burner will experi-

ence the anguish of watching a fire wither and die. If he knows that his experience is shared by every neophyte coal burner, however, he will be more likely to try again than to blame his problems on the stove.

We start at the beginning, telling the new stoveowner to clean off the protective oil on the griddle before lighting the first fire. We explain the proper procedure for breaking in a cast iron stove. Then the presentation moves on to a simple explanation of combustion principles and how they relate to the various features of the stove. Finally, we delve into the responsibility that goes along with stove ownership and the commitment to routine maintenance that spells the difference between merely using a stove and achieving the levels of performance and service of which a good stove is capable.

Any Questions?

Stove School takes as little as ten minutes or as long as several years, depending on the curiosity of the customer. For the staff member who graduates as many as a dozen classes on a busy fall day, the experience will be as grueling as it is exhilarating. Yes, creosote has been explained twelve times to twelve different groups, but there is satisfaction in knowing that you have helped twelve people avoid what could have been an unpleasant surprise. Ultimately Stove School helps stoveowners achieve satisfaction with their new purchases. And having satisfied customers is good business.

Not everyone buys a Vermont Castings' stove, and not everyone comes to Randolph, Vermont. But the basic curriculum is valid for all stoves. We have learned the hard way. Let us save you the trouble. There is one prerequisite, though, and that is owning a stove. Before we start Stove School officially, let's consider what you will want to know before you purchase a stove.

Apples for Cider

A Buyer's Guide

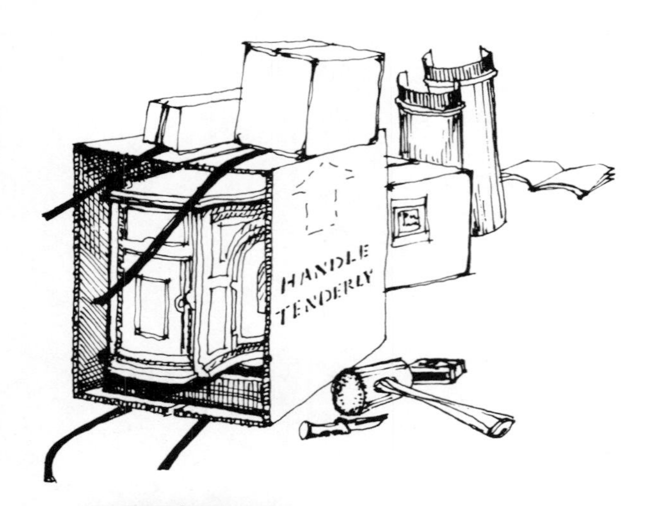

Choosing the correct stove is a little like choosing a mate for life. It is something you will have to live with on a daily basis. There is a bewildering array of styles and sizes to choose from, and you will be unhappy if you make the wrong choice. And, like choosing a mate, you will know for sure if you have made the correct choice after spending a long, cold winter with one. Fortunately, buying a stove is usually an easier and more tangible process.

Most people who buy a Vermont Castings' stove have done a great deal of homework before making the final decision. Many started their research by perusing the "Woodstove Directory," an annual publication put out by Energy Communications Press, Inc. of Manchester, New Hampshire. Basic information on popular stove models is available in this magazine, as well as a product locator guide to let you know where these products may be purchased. The descriptions of the products in this directory are provided by the individual manufacturers and will present the respective products from the most attractive perspective. They will all sound good, and you will probably want to see one to better evaluate the quality.

Where to Buy

Stoves are available in stores specializing in hardware or building supplies, large chain stores, stove specialty shops, and direct from the factory. The last two offer the best bargains in quality and service.

Factory Direct. Buying from the factory will usually allow you to get the best price, since no middleman is involved. The factory will also be the most knowledgeable source of information of the product. You should appreciate the "do-it-yourself" ethic when buying from the factory, however, since you will have to arrange for delivery of the stove and install it yourself. The reputable manufacturer or retailer will be a member of the Wood Heating Alliance, the national trade organization for the stove industry.

Stove Specialty Shops. A number of quality shops have emerged across the country to meet the need for stoves generated by the energy crisis. Many of these shops provide services such as delivery, installation and chimney cleaning, or can put you in touch with reliable local alternatives. The price of the stove will usually be higher, but the availability of their service will probably make it worth it.

The salesperson should ask you lots of questions about your house, the floor plan, and your heating expectations. This indicates a sincere desire to help you choose the right stove for your home rather than the one with the highest profit margin for the shop. Ask the salesperson about his or her experience with burning wood or coal. They will need to have had some experience themselves in order to help you.

Finally, ask for references. A quality stove shop should have no trouble providing you with a list of happy customers. (*Note:* A listing of some reputable and recommended stove shops will be found at the back of this book.)

The Fine Print

Read the warranty on the stove carefully. You have a legal right to request to see the written warranty if it is not immediately available. Don't be fooled by what seem to be multi-year assurances. A twenty-five-year warranty may be qualified to such an extent that few problems you are ever likely to have will be covered. Also, ask the salesperson about parts inventory during the heating season. Any part you are likely to need should be readily available all year long.

Craftsmanship

It is always interesting to read the literature of the various stove manufacturers, but somewhat confusing if you're trying to gain insight as to how well any particular unit may be constructed. There can be a large discrepancy in the quality of stoves within any of the three major material categories used in their manufacture: cast iron, soapstone, and steel plate. How do you tell if one stove is better than another? Hopefully, we can answer this question by providing an informal checklist that you may use to judge their relative qualities. None of the points alone guarantees that the stove is a good one, yet each provides some measure.

For most people, craftsmanship represents how well a product is made; that is, how well the design has been executed. Often, however, it is difficult to separate design from execution. What we often interpret to be poor quality may actually be poor design and not shoddy workmanship. The result is the same—an inferior product.

There are many visions that personify the term "craftsmanship," but the word often conjures up thoughts of an old cabinetmaker at work in his warm and cluttered shop, carefully fitting the last dovetail joints on the drawer of a magnificent Goddard Townsend block-front desk. Nostalgic, perhaps, yet a scene such as this is where the ideal of craftsmanship becomes clear for many. It is not so much a matter of specific skills, or attention to detail, but rather it is an attitude. The craftsman has a sense of purpose to create items that exhibit both beauty and function. From this attitude come the tangible details that may be recognized as examples of craftsmanship.

THIS CAST IRON LOOKS SOFT AS A SOFA IN THE

Cushion 25

1861

Building a stove is much like constructing a piece of furniture in that the design is dependent upon the materials used in its construction. Each type of wood has its own unique set of properties that dictate design and construction details, and the same applies to cast iron, soapstone, and sheet steel.

Cast Iron

Cast iron is the most fluid of the three stove-building materials. It is produced by sand casting, a process by which patterns are embedded in moistened sand and removed, leaving a cavity into which molten iron is poured. This is a highly technical process involving an entire industry of related skills from pattern-making to metallurgy. Because it begins in a relatively liquid state, cast iron can be formed into virtually any shape. Cast iron stoves generally exhibit the widest selection of styles and ornateness. With a material that is so readily shaped, designers today as in the past are offered unique artistic opportunities.

Quality castings are characterized by a consistently fine, grainy surface free of blemishes, and with the detail reproduced crisply. Generally, any castings of lower metallurgical quality that exhibit blemishes are susceptible to breakage or warping. A standard wall thickness of at least 3/16 inch is necessary to provide sufficient strength to each part.

Cast iron stoves are usually made up of several plates that interlock and bolt together. This construction technique allows the individual stove plates to be replaced should any damage ever occur.

The stove body usually has six pieces: two ends, and a top, bottom, back, and front that join together, sometimes with tongue and groove joints. Seams and joints are sealed with stove cement to assure airtightness. Some stoves have interior tabs that mate and are bolted; others use metal rods that bolt the top and bottom together, holding the sides, front, and back in place. The important thing to note is how solidly the stove is held together. If there is any movement at all, the stove is probably not airtight.

One advantage of cast iron is the ease with which it can hide joint lines behind moldings by using overlap construction. Stove manufacturers who take care to do this usually have spent time fitting pieces to assure that it works well. The one danger in using moldings in this manner is that any misalignment is quickly visible. However, the resulting good fit provides an exterior on which the panels work as an integral unit, and not just a box made up of six pieces.

Soapstone

Soapstone stoves are an integration of cast iron and soapstone. While the cast parts can be judged by the standards mentioned above, the soapstone panels that give the stove its unique qualities are subject to other considerations. Soapstone is a natural rock formation (steatite), a form of shale. It is mined in blocks from quarries and sliced into sheets that become panels for the stoves. The relatively soft stone is finished by means of sanding and polishing, and since hardness and coloration vary depending upon the quarry, each panel exhibits a marble-like grain slightly different from the next.

The surface of these panels should be consistently smooth and flat with no surface marks. The

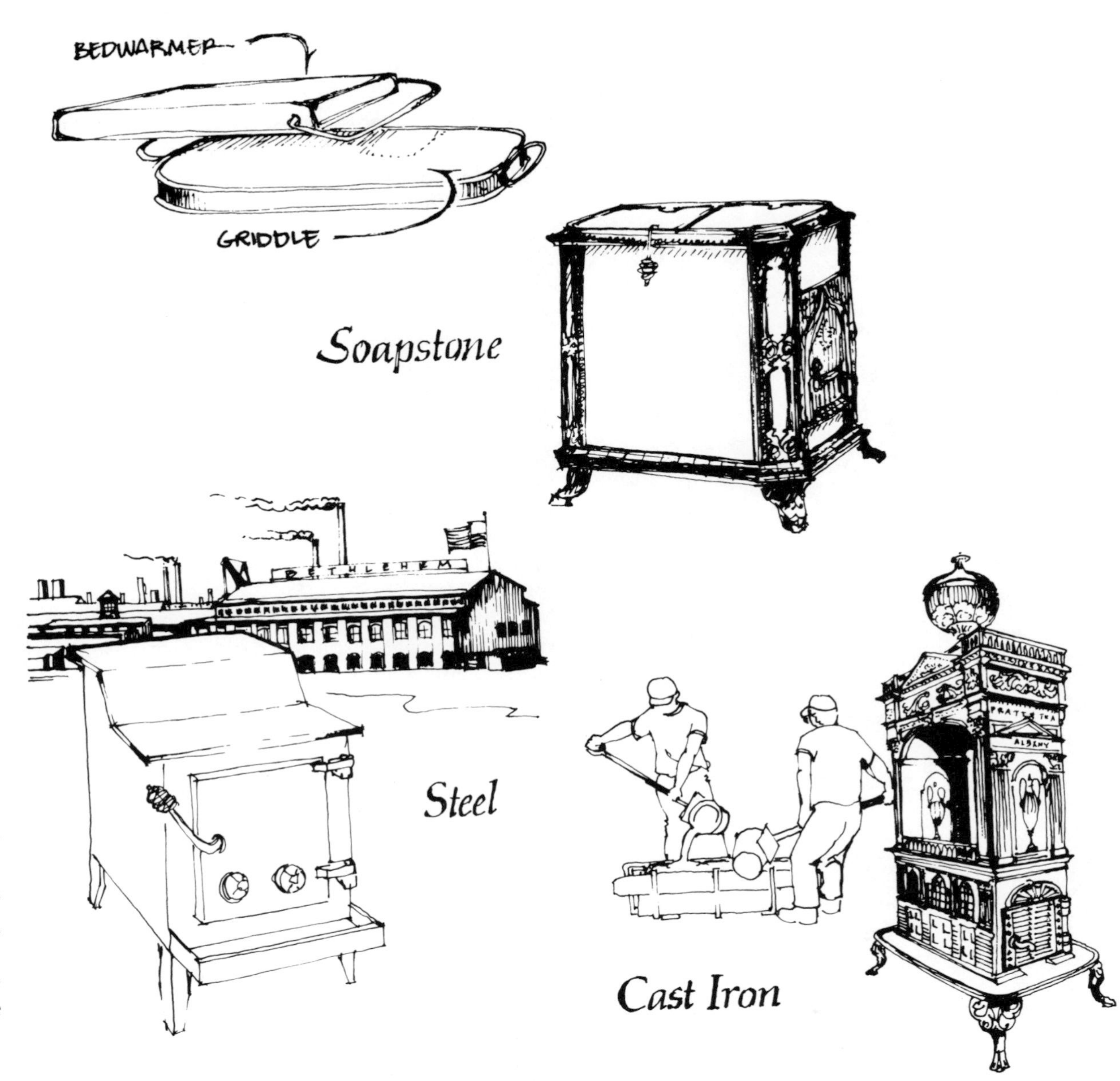

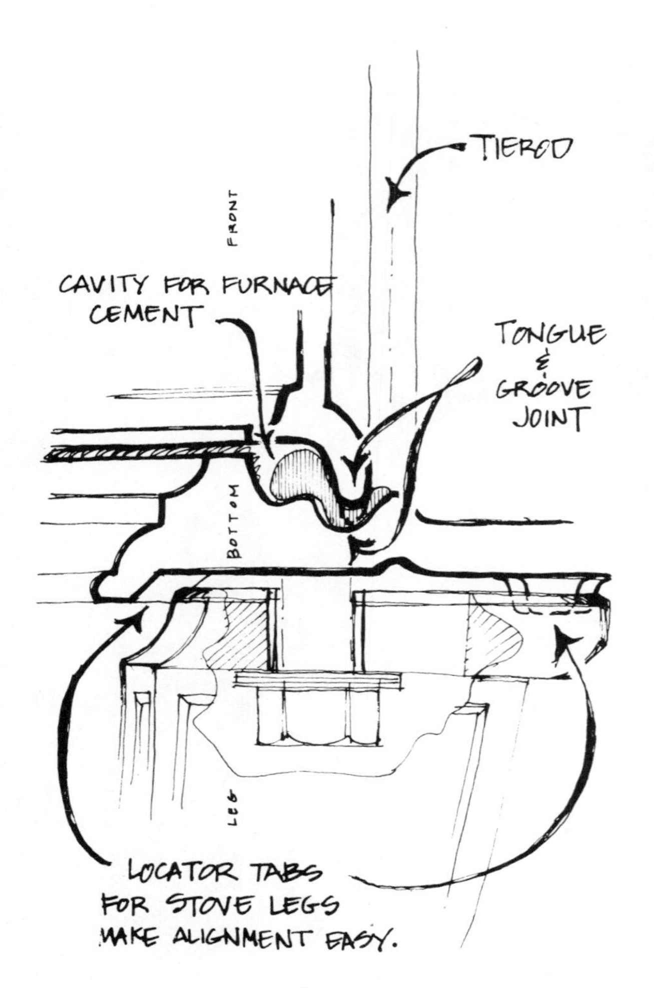

Stove Details

panels should be sufficiently thick (¾-inch) for suitable strength.

Large, flat panels are the most common use of the blocks that are quarried, but designs need not be limited to either flat surfaces or large panels. The color and grain of the soapstone chosen during the construction of the stove will affect the overall appearance. Expect the same care in choice of the soapstone panels that you would expect from a cabinetmaker in the choice of wood grain for furniture panels.

The construction of soapstone stoves features a cast iron frame that joins and holds all the panels together. The panels are held in the frame by a system of slots or grooves. In some cases, the soapstone itself is machined to secure the pieces to one another. This system allows an overlapping of the panels, giving the stove a clean, neat appearance. Airtightness is assured by using gasketing and cement between the panels and the frames.

One stove on the market uses a series of mini-panels glued and splined together to create the larger panel for the front, back, ends, and top. This creates an interesting visual affect, but the greater joint area increases the chance for air leakage. The smaller panels are cheaper than large ones to replace, however, in the event of damage. Given that the frames are bolted together and easily disassembled, all soapstone stoves are generally rebuildable.

Steel

Steel stoves are made of panels cut from sheet steel or boiler plate, typically 1/4 to 5/16 inches in thickness; the sheets offer good resistance to warpage under extreme temperatures. The appearance of steel stoves is generally box-like, with few stylistic variations other than simple bends or step tops. Usually, stylistic accents come in the form of add-on ornaments like birds, or scenes that are executed on cast iron doors.

Steel does have the advantage that it can be welded. Welding not only provides for sturdy construction of the stove, but also makes it permanently airtight as well. The quality of the welds is important; they should be straight, smooth and continuous. Spotty welds should be suspect. All corners should be smooth and edges rounded to be safe. The plane of the steel sheets should be straight; that is how steel plate is manufactured, and any variation is an indication of poor workmanship. The pieces should fit squarely together.

Ideally, the stove is constructed so that as few weld joints as possible are in view. Some manufacturers manage to hide most of the joints by using bends and overlaps, but often the lack of moldings and other decorative details prevents all from being hidden.

Beyond Materials

There are other aspects of stove design that should be taken into account; one of them is legs. Stove legs should be able to support the great mass above them. They need to be high enough to vacuum or sweep under and, if possible, should provide some means of adjustment for uneven hearths. Legs also should be high enough to meet any local codes requiring that the unit be a certain distance off the floor. Some legs are permanently attached by welding. The rest must be attached manually. A leg attachment system that utilizes a bolt and some locating tabs for proper leg alignment works well. The bolt itself needs to be substantial, for it is carrying the load.

The door system of a woodstove is the most handled and most mechanically complicated part of any unit. In many cases, it is the only moving part on the stove. Regardless of the stove type, most doors are cast iron. Even on steel stoves, cast iron doors are used to insure good fit since the iron has less tendency to warp than steel. Iron makes it possible to add decorative flair to the front face of an otherwise plain stove. No matter what the configuration of the opening, a cast iron door can be produced that will insure accurate sealing. This seal is important. For many stoves (steel plate in particular), this is the only opening through which air might leak.

A well-designed door system should include a high temperature gasket to insure a good seal. If there is more than one door, they need to overlap in order to seal the vertical seam that they create. This seam is usually the hardest on which to maintain a tight seal, and should be gasketed as well.

Hinges allow the doors to move freely and accurately. Usually, half of the hinge is attached to the front stove plate and the other half is part of the door itself. Accurate design, drilling, and assembly of the hinges is vital. Improper positioning of hinges can lead to bad sealing of the door, or can lead to doors that bind when you open them. It is not only annoying, but dangerous for the door on your wood stove to malfunction in any way. The care that the manufacturer takes in fitting these will tell you a great deal about the stove's quality.

The last system on the doors is the handle-latching system, without which the door would open the first time a log fell against it. The latch system pulls the door or doors together and against the front, securing them from opening while the fire is going. The latch engagement should be good and tight enough to bring the doors firmly against the front. The latch system should have a built-in adjustment to allow for wear of the gasketing; otherwise the stove's air-tightness will be short-lived.

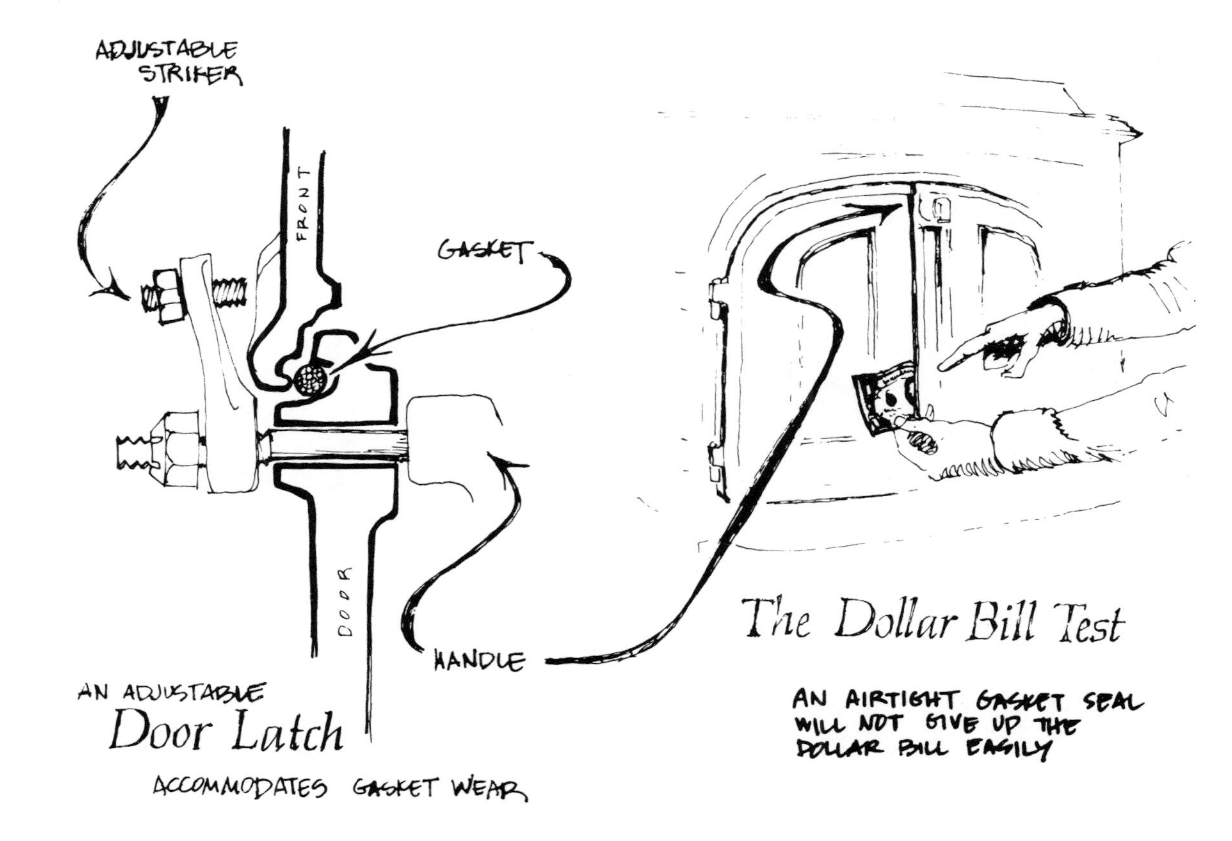

Learn from the "Experts"

Fully understanding the operation of your stove will allow you to achieve maximum efficiency from it and may prevent potentially dangerous or embarrassing situations.

Man and Burning Log

"That first winter in Vermont we were living in a basement apartment heated with only a Vigilant stove. My wife, new to woodburning, woke up late one night and decided to check the stove. Forgetting to first open the damper, she opened the front doors, allowing a partially burned log to get stuck half-in and half-out of the stove. Within a few short moments, air rushing into the stove had fanned the flames enough to turn the stovepipe glowing red. At this point she awakened me. Quickly sizing up the situation, I donned a pair of heavy leather ski gloves and shouted "Open the door!!" I pulled the smoldering log from the stove, dashed past my wife, and out the door. In addition to the ski gloves I was wearing . . . well, I wasn't wearing anything.

Standing outdoors in the 20° below zero weather with the danger past, the absurdity of the scene struck me—a man clad only in ski gloves, arms extended upward with a trail of sparks arching across the starlit Vermont sky in the middle of the night. An observer would have thought me a celebrant in some strange pagan ritual."

Stephen Morris

Wood stoves at some time became parlor stoves, a label apparently inherited from their acceptance in the house as a part of the general interior furnishings. This transition to an attitude that began to think of a stove as a piece of furniture is important. A stove should look as if it belongs in the house and not just in the basement. It should, as should any valued furnishing, add grace and beauty to the room and reveal the thought and care that went into its design. Taste in design is, for the most part, purely subjective, but it is important to question if the manufacturer of a stove is as concerned with craftsmanship as with production.

Other Features To Look For

There are many other features and options that help to distinguish some stoves from the rest. Some of these characteristics are desirable and others are not. The most popular ones seem to be the following:

Dual-fuel Burning Capability. Being able to utilize either coal *or* wood as fuel should be a consideration unless you have an unlimited and inexpensive source of one or the other. Don't be fooled by claims that either fuel can be burned as effectively within the same stove without modifications.

Each of these fuels requires a special combustion set-up. Vermont Castings' parlor stoves, for example, solve this problem by incorporating into the existing stove a kit designed especially for burning coal. This essentially makes a "stove within a stove."

Water Heating Ability. Some stoves allow you to hook your water supply system to your stove so that the stove heats your water. The installation of the pipes requires a thorough understanding of hot water system requirements and should be done by a professional. Serious damage or injury could result from improperly installed systems.

Stove Top Cooking. The surface should be flat and able to accommodate one or more cooking utensils. The pleasure of "back-burner" cooking will become quickly apparent during your first winter.

Fire Viewing. This may be provided by a special, heat resistant glass for either wood or coal stoves, or by opening the door(s)—on a wood stove only. The glass has a tendency to fog up when wood is burned, but for the most part it is self-cleaning.

Catalytic Combustors. These devices, once perfected, will allow stoves to burn more cleanly and efficiently. To date the combustors work only sporadically. They are easily contaminated by combustion impurities and require regular replacement—an expensive proposition. Ask other combustor users about their experience before making this decision.

Price Range

You get what you pay for. You can spend under $30.00 for a small tin box or barrel stove (you supply the barrel) or more than $2000.00 for an elaborate cook stove. A quality stove will be a lifetime investment, so don't compromise on the initial purchase price.

Sizing a Stove to Your Needs

The final judgment in the selection of a wood or coal stove is perhaps the most important: The stove should be sized to your heating requirements. The temptation of many American consumers is to buy a product that is "more than enough." This attitude has filled our lives with blenders that have more speeds than we can ever hope to use and automobile engines capable of moving the vehicle to the unlikely speed of 120 miles per hour.

A woodstove is one investment to which the "biggest is best" rule does not apply. A stove that produces more heat than you need will push temperatures in the immediate area to uncomfortable levels. When the stove is damped down in an effort to control the heat output, the fire smolders inefficiently and creosote becomes a problem. Conversely, a stove that is too small will leave you cold and unhappy. You will be tempted to fire it beyond its capacity, resulting in a shortened length of burn and possible damage to the stove.

The following checklist summarizes some of the basic considerations of stove size. It is not meant to be a scientific selection guide, but rather a general indicator of how a certain stove size relates to certain needs. Assign points to each choice according to the following schedule:

a. 20 points
b. 15 points
c. 10 points
d. 5 points

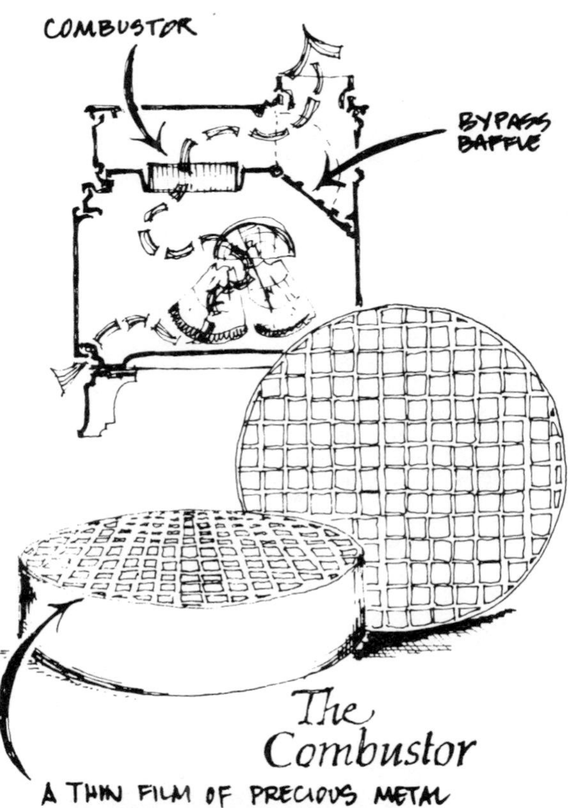

1. *Size of area to be heated*
   a. *8–10,000 cubic feet*
   b. *7–9,000 cubic feet*
   c. *6,500–7,500 cubic feet*
   d. *Less than 6,500 cubic feet*
2. *Expectation of heating capacity of wood/coal stove*
   a. *Sole heat source*
   b. *Primary heat source*
   c. *Supplemental heat source*
   d. *Not a part of home heating plan*
3. *Number of rooms in area to be heated*
   a. *More than 4*
   b. *3–4*
   c. *2*
   d. *1*
4. *Total Heating Degree Day Units for your area*
   a. *8000 or more*
   b. *6000–8000*
   c. *3000–6000*
   d. *Less than 3000*
5. *House efficiency rating*
   a. *Little, if any insulation, drafty*
   b. *An older home; some insulation has been added*
   c. *Home built since 1970 with standard insulation*
   d. *Very tight, well insulated*
6. *Where stove will be located*
   a. *Basement installation*
   b. *Center of large open area*
   c. *On outside wall of main floor*
   d. *In separate addition to home*

You can now get an approximate idea of how large your stove should be in terms of its BTU output:

Points	*BTU Range*
96–120	55,000 or more
65–95	40,000–55,000
36–65	30,000–40,000
35 or less	less than 30,000

Welcome to Stove School

By now you have satisfied the prerequisites for this section; you own a stove and have it installed, or you are close to getting one and have given some thought of where you will put it in your house. The goal of stove school, as with any good educational experience, is to provide you with basic information as well as to inspire you to ongoing independent learning. The basics have to do with stove operation and maintenance, and the independent learning will take place as you gradually master the art of wood and coal heat as it applies to your own home.

Elementary Woodburning

Every new woodburner starts his career with a first fire. He often has picked up information in preparation for the event from a variety of sources, including the salesperson from whom the stove was purchased, the operation manual, neighbors who are woodburners, and hopefully from a thorough reading of *The Book of Heat*. He has determined what stove preparations are necessary before the initial fire can be lit. Some stoves, for example, require that the bottom of the firebox be covered with sand or wood ashes as insulation to prevent thermal stress on the new metal.

Many new woodburners choose to have the first fire or two with the stove sitting in the middle of the backyard. This practice is not as absurd as it may appear. Like the new muffler in your car, new stoves often produce an unpleasant smell when they get hot for the first time. Many new stoves are finished with a high temperature paint that will smell when heated. The resulting fumes are not hazardous, but they are not pleasant either. In addition, new metal has an odor when first heated. Fumes from the "curing" of a recently fired stove can occasionally set off your smoke alarm, so if you do not have the first fire or two with the stove outdoors you may want to choose days when you will not mind opening the window.

Once the stove is finally established at its permanent home on the hearth, it is not uncommon for the new stoveowner to discover that he is afraid of it. The fear is not so much of the stove, but of having hot combustion taking place in a metal box right in the parlor rather than being safely hidden in the basement, out of sight and out of mind. Few woodburners will confess this

fear to their more experienced neighbors, nor will the experienced neighbors ever offer that they have felt the same. Everyone experiences the feeling of uneasiness to some extent, though, and it is a good thing. Out of this initial fear grows the cautious respect that is essential to a healthy and safe relationship between the stove and its tender.

The term "break in" is an unusual one, meaning in different applications "to teach" or "to condition." We break in a new employee, or we break in a new pair of shoes. In both cases there is an implication of getting used to or becoming accustomed to something. New wood stoves need to be broken in as well, for two reasons. The first is to allow you to learn the controls and feel comfortable with the stove, and the second is to allow the metal the opportunity to expand and to contract gradually and to introduce the metal to heat.

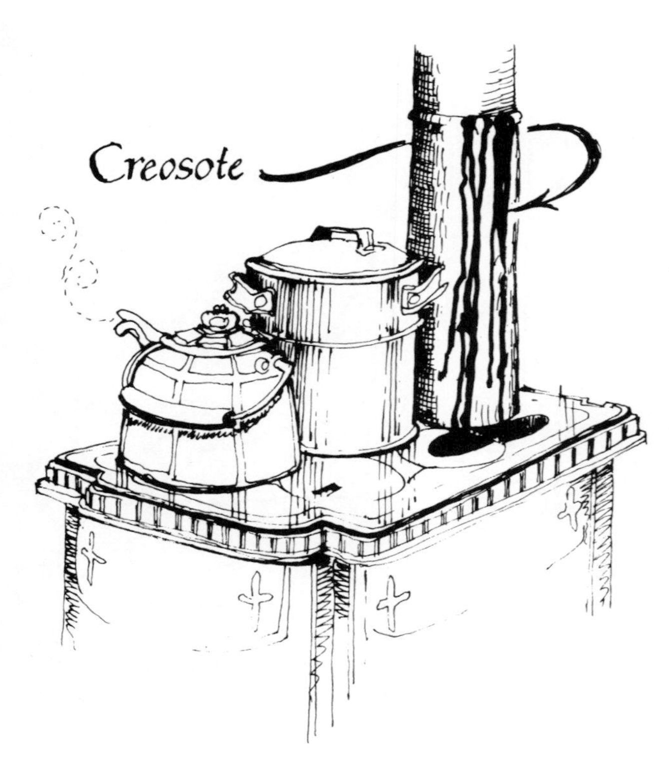

Plan to have a half-dozen break-in fires in your stove. Open all air inlets fully. Start with light kindling and newspaper, and add finger-sized pieces until the fire seems to be going independently. Next add one or two wrist-sized sticks, and a couple more in ten minutes. Plan to spend up to an hour gradually building the fire, until the fire is established and the stove is hot to the touch. Maintain the heat at this level for the next 3 or 4 hours, then let the fire go *cold*. This is your first break-in fire. You will want to have five more before putting the stove into full heating service. (Some stoves may not require a break-in procedure, but start gradually anyway.)

Do not close off the air supply completely during the break-in procedure. Since the operating temperature of the stove is already being kept at a low level to prevent thermal shock, an attempt to restrict the air completely will encourage the flue gases to cool and condense as liquid creosote. A moderate but crisp blaze is what you want to keep the flue clean. Many new woodburners start a break-in fire in the late afternoon or early evening. By bedtime the fire is still going, and the inclination is to "shut it down" by closing off all air. Sometime during the middle of the night or the following morning, the new stoveowner may awaken to an acrid smell and may find hardened black rivulets on the stovepipe or wall. This will not happen if the fire is allowed to burn crisply.

Fires during the break-in period prefer wood that is finely split. The thinner pieces of wood burn more easily at the low combustion temperatures for which you will be striving. Large chunks will have a tendency to smolder, making it difficult to sustain the heat level you want and making the stove seem sluggish. A smoldering fire is one that encourages creosote.

Effects of the Weather

One advantage of stove ownership is that the commitment makes us more aware of the world around us, particularly the weather. Many stoves are put into use for the first time in the fall or in the spring, just before or right after a winter of high fuel bills. Since chimney drafts are often lower during warm weather, novice woodburners sometimes begin learning about their new heater under the worst possible situation.

The low draft experience most frequently happens with stoves that allow the front door to be opened for fire-viewing. An open fire requires plenty of oxygen and a strong draft. The open door also provides a place for smoke to escape, and the chimney that allows the stove to burn pleasantly all during the winter with the door open may allow the stove to smoke in early spring and late fall when outside temperatures are warm. Fortunately, a warm-weather smoking problem can be corrected fairly easily. If the smoking stove has only one door, close it, and fully open the air supply to the stove for a few minutes. If the stove has double front doors, close one of them. Either method should increase the rate of combustion, which will cause the flue to become warmed by the hot gases entering it, and your draft will be re-established.

You can then open the door again. This procedure should work in most cases, but may need to be repeated several times in the course of an evening.

An opposite draft situation can occur during a winter cold spell. When the temperature drops well below the freezing point, your chimney draft can improve to the point where it is causing undesirable performance changes in the stove. Symptoms of unusually high draft include shorter than usual burn time, slower rate of response, and a red glowing of the stove or smoke pipe. These symptoms may all be present at the same time. The solution is as simple as reducing the supply of incoming air, but this requires caution when the stove is very hot. A considerable quantity of volatile gases will continue to be given off by the fuel, and the stove can become congested with volatiles. When incoming air is mixed with the gases a very rapid burn can occur and result in "backpuffing." Always reduce the air supply gradually to a very hot stove.

Wind can also be a factor in stove performance, particularly if your house location is geographically extreme. Your chimney will be most susceptible to the wind if you live near a body of water, on a high hill, or in a low valley surrounded by high hills.

A Chimney Cap

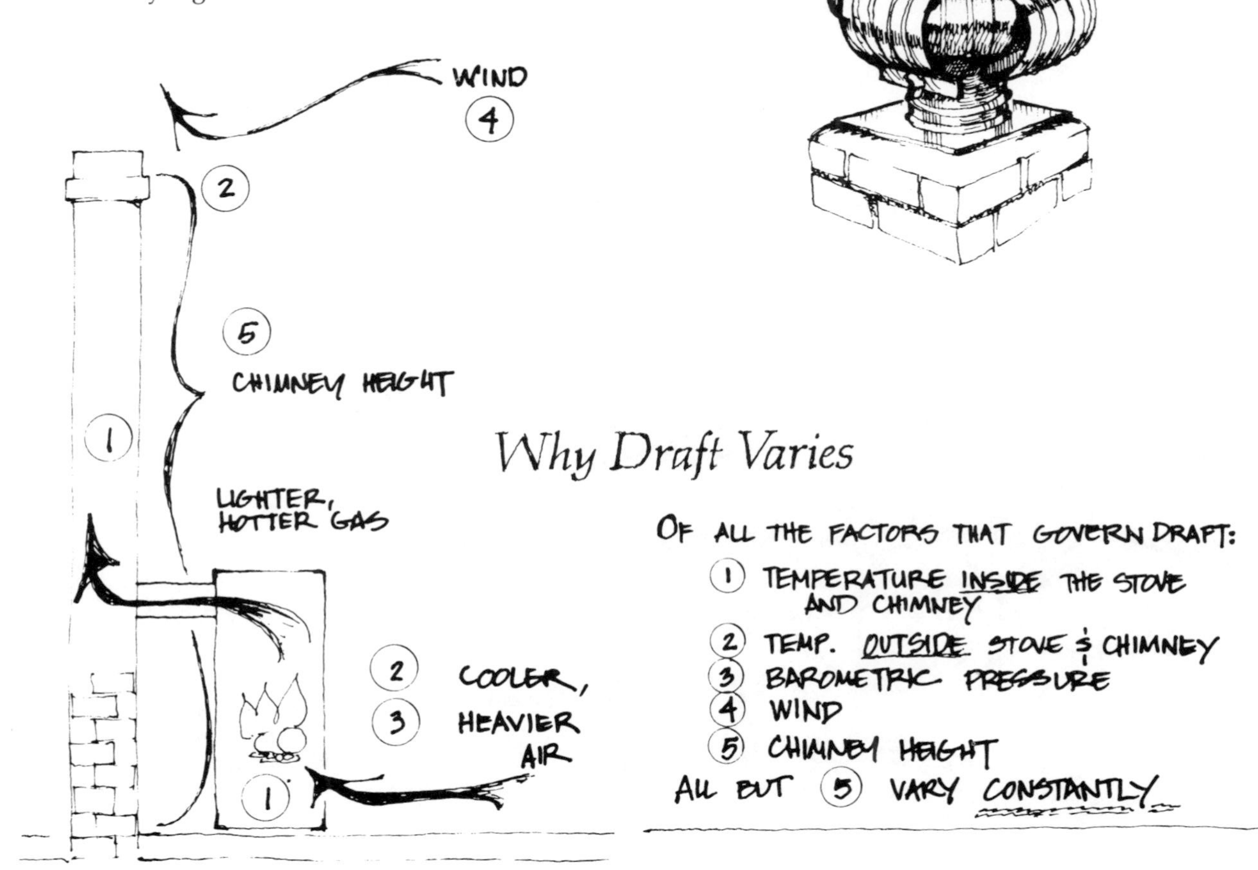

The wind generally creates a draft problem when the stove is being burned at low levels. The low burn rate sets the stage by sending little heat up the flue and allowing it to cool. Without the upward pressure of the hot, buoyant gases the chimney can become an avenue for "down-drafts," especially if the wind is gusting. The downdrafts can force smoke out of the stove in the house, and create backpuffing, or small muffled explosions within the stove. Backpuffing from the wind can be frightening, and some geographical locations are much more prone to it than others. Wherever you live, there are several things you can do about it. Maintaining a higher stack temperature is the best cure. Burning your stove at a higher temperature will send more heat into the flue and result in a more positive upward draft. A quick solution can be achieved by opening the internal damper of your stove if it has one. This immediately sends more heat up the flue.

A more expensive and possibly more permanent solution is available by increasing the height of your chimney. It is not predictable that increasing the height will work, but if it does it should be a permanent cure.

The third remedy is to install a chimney cap. As with the chimney extension, there is no guarantee that the cap will work; if it does, it should be a permanent solution. There are a variety of caps on the market that will help to stabilize draft under windy conditions; there is still no firm evidence that any of the caps will actually improve what is already a weak draft.

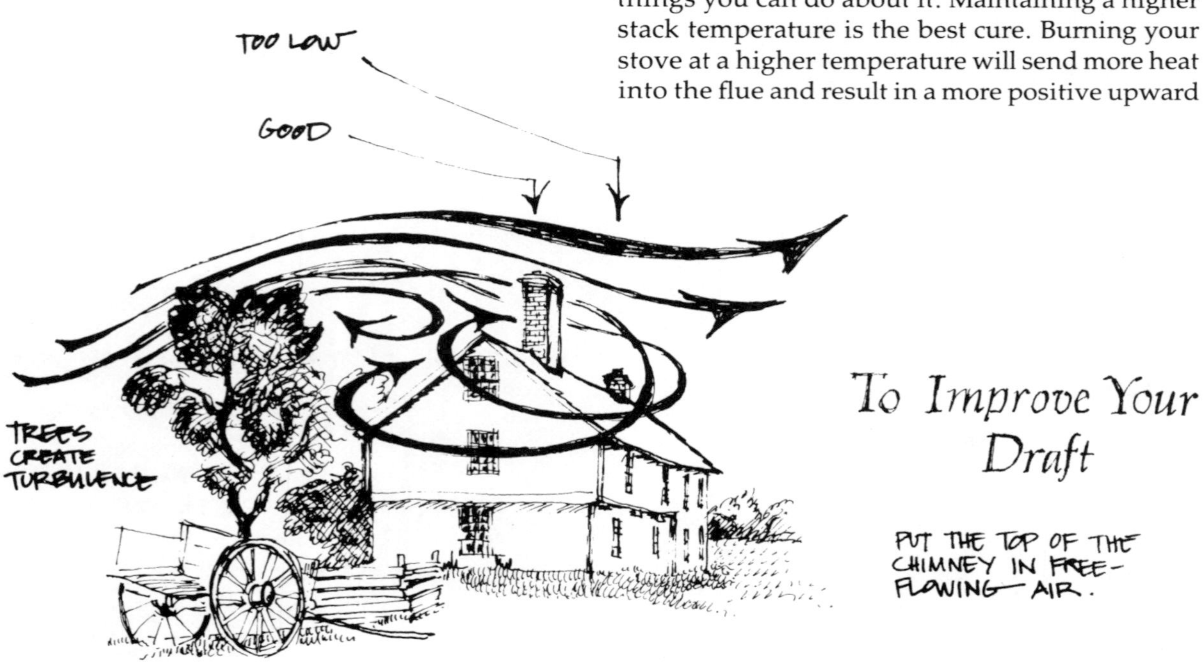

To Improve Your Draft

Coal Burning for Beginners

Coal burning is an elective course in Stove School. It offers special challenges and requires that you fully understand some special techniques. The rewards will be great, however. The successful application of coal-burning wisdom to your home heating plan will keep you comfortable through the winter and save money on your fuel bill. But first, you will have to earn it.

Few things give rise to panic as much as an unsuccessful attempt to get your first coal fire started. If you are a typical new coalburner, you have spent several months comparing the numerous models available before settling on your stove. You probably have two or three tons of precious fuel safely stored downstairs in the cellar, and you might have worked hard to convince your mate that a thousand dollar investment in a stove and its installation will quickly pay for itself in fuel savings and the pleasures of self-sufficient heating.

While coal as a fuel offers some advantages not available with wood, it is also more demanding. Initial attempts at starting or keeping a coal fire going can be disappointing if careful attention is not given to crucial steps in the procedure. Here are some tips that will help ensure your beginning success in burning coal:

Start-Up Procedure

1. Don't rush your start-up fire. This is a procedure that does not need to be duplicated regularly, so do it right the first time.

2. Use dry kindling, and be generous with it. Start with crumpled newspapers and large slivers of softwood. Gradually add chunks of hardwood kindling until there is a good thick bed of red embers covering the grates.

A variation of this starting technique is to use charcoal briquets. (Unlike a fireplace, an airtight coal stove should allow no charcoal fumes to escape.) Place two to three pounds of briquets directly on the newspaper, and ignite them in the same fashion you would ignite your summer barbecue charcoal. Whether using wood or charcoal, you might find that cracking a front door slightly open will help. (***WARNING***: never use flammable fluid to start the fire.)

Whichever technique you employ, the kindling procedure might take an hour or more and helps to preheat the flue as well as to provide a well-established fire base on which to place the coal.

3. Add the coal gradually. Sprinkle two or three shovelfuls evenly on the kindling base. When this has ignited sprinkle on another shovelful or two, and continue in this fashion until the coal has become part of the kindling base. If possible, use pea coal at first, then add your nut coal. If you do not have pea coal, add the nut coal gradually as described above.

4. If your stove has a magazine, fill it only after glowing coal covers the entire grate area and the firemass reaches up to the bottom of the throat.

5. Close your air supply only after the fire is burning well. If you do not yet have a thermometer, you should get one; it will be well worth the money in helping you to monitor your stove.

In on-going operation of the stove avoid letting the surface temperature drop below 200°. Coal fires below that level are difficult to revive and may die out completely.

Shakedown Procedure

The coal ash that accumulates as the coal burns must be removed from the grates regularly for the fire to continue at a comfortable level. Coal stoves generally incorporate one of two methods: The first is the "shakedown" approach. An externally

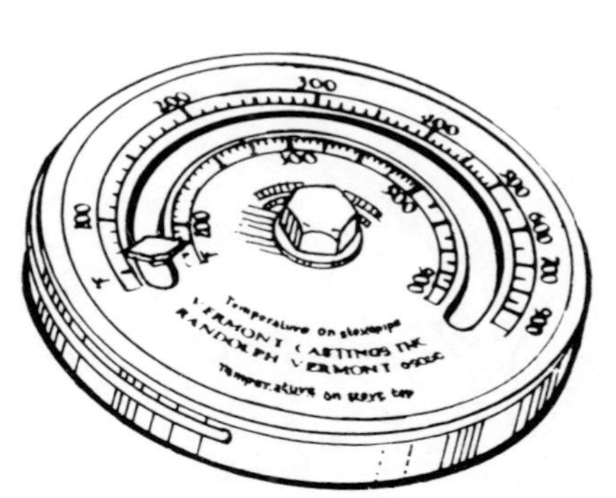

A Stove Thermometer

mounted handle allows the operator to "shake" or rotate the internal grates so that the ash will fall from them. The second approach employs a "slicer," a tool that resembles an oversized butter knife, to dislodge the ash manually. The ideal stove design will incorporate both methods.

Whichever method your stove uses, the purpose is to remove as much dead ash as possible from the grates so that the fresh supply of coal will burn freely. Terminate the procedure when incandescent coals begin to drop through the grates; you don't want to dislodge the entire hot bed of embers.

Don't be tempted to shake or slice down the ash too frequently. Exact frequency will depend on stove type, but generally, higher operating levels will necessitate more frequent ash removal.

A Word to the Wise

Coal and wood require different stove designs in order to burn effectively. Claims that a stove can burn either fuel with equal effectiveness without some modification of the stove should be carefully investigated. Some manufacturers have solved the dual fuel problem by designing special kits that can be used for burning coal. Vermont Castings' coal stoves, for example, incorporate an unusual design concept known as the "stove within a stove." Just as it is important to have an airtight seal between the exterior plates of the stove, it is also crucial that the "inner stove" be sealed well in the correct areas for proper performance.

This cementing procedure will be required when you are installing a new coal kit into a stove previously used for burning wood or if you are converting your coal stove back to coal burning after having removed the "inner stove" in order to burn wood. Occasionally, the cemented seals may develop leaks from the expansion and contraction of the heating/cooling process as well. Always be generous with the stove cement when resealing the coal-burning components, and follow the directions in the manual closely. Be sure to cement the left, back, and right grates, the bottom exhaust manifold, and the exhaust cover on Vermont Castings' coal kits.

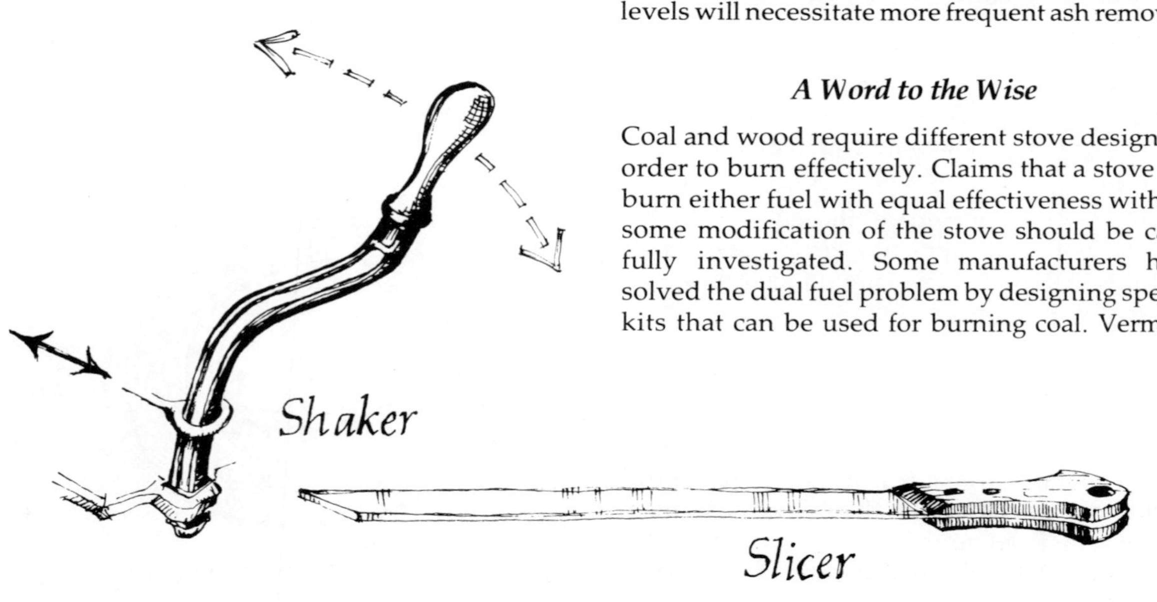

About Draft

A coal stove needs an adequate draft in order to operate properly. If your chimney does not provide a good draft, your stove will not perform as well as it should. It is helpful to understand some basic principles of draft in order to correctly analyze a draft-related problem. The most important fact is that hot air rises, and the greater the difference in temperature between the air in your flue and the air outside, the faster it will rise. Cool air enters your stove, stimulates the burning fuel, and leaves the stove at a much higher temperature. Since it is warmer than the surrounding air, it readily rises up the chimney. Draft suffers if the air rising up the chimney cools.

If your chimney is on the outside of your house, for example, it will cool rapidly. As the chimney cools, the air and gases within it cool as well. The cooler they get, the less eager they are to rise. The same can happen if your flue area is too large. The warm air leaving the stove is so diluted by the cool air in the large flue that it quickly loses warmth and buoyancy. Large fireplace flues and poor fireplace connections can create a similar problem. Fortunately, there are some things you can do to compensate for a poor draft. Below are some suggestions:

1. *Preheat the flue.* If inadequate draft hinders your attempts to get the coal burning, it will help to preheat the flue. Preheating is accomplished by having a lengthy kindling fire to supply the flue with enough heat.

2. *Burn your stove at higher temperature levels.* Usually a temperature in excess of 500° is sufficient. This insures that the air and gases entering the flue will be well warmed. (*Note:* This procedure might require some creative ways to utilize the additional heat that will be generated into the room.)

3. *Extend the stovepipe into the flue.* This is particularly true of fireplace installations; extend the pipe past the damper as much as possible, but at least a foot into the tile. It is essential that there be an airtight seal between both the pipe and the damper plate, and the plate and the flue.

A Stoveowner's Vocabulary

No subject can be fully understood without an understanding of the key words. The following glossary covers some of the key terms used to describe elements of stove design and combustion.

Airtight. A stove designed and built in such a manner that essentially all air entering the stove passes through one or more controllable air inlets. No stove is truly "airtight;" a more definitive term would be "air-controlled."

Ash. The inorganic compounds remaining after combustion of a solid fuel. In the case of wood, ash is comprised of the trace minerals that were once part of the make-up of the tree.

Backpuff. The result of the rapid burning of smoke and volatile gases within the firebox, evidenced by smoke and sometimes flames being forced back out of the stove. This is most likely to occur with a smoldering fire when a small quantity of fresh air is suddenly allowed to enter the firebox.

Baffle. A metal or ceramic partition designed to direct incoming air or exhaust gases through a specific path within a stove. A baffle can increase heat exchange by forming a longer path with more surface area, or cause a stove to burn with certain characteristics as a result of the way in which incoming combustion air is directed, e.g., Scandinavian baffle.

BTU (British Thermal Unit). A quantitative measure of heat. The amount of heat needed to raise

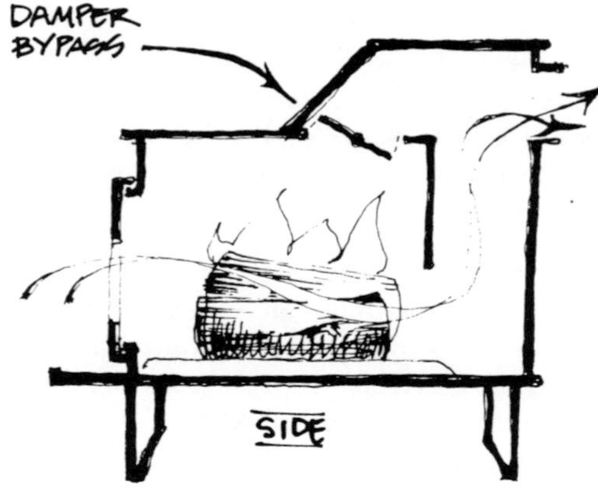

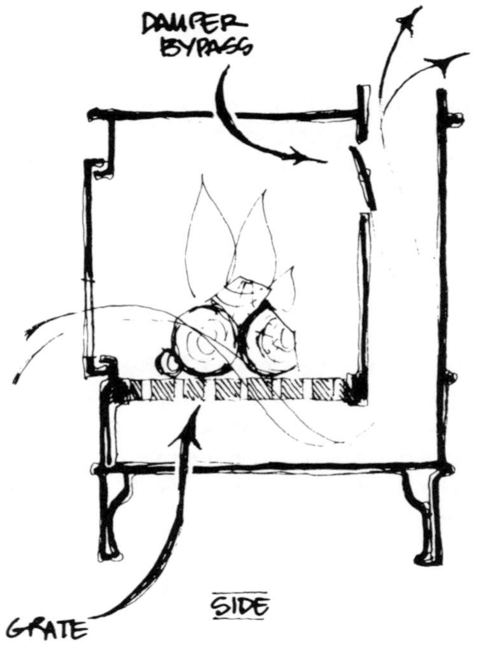

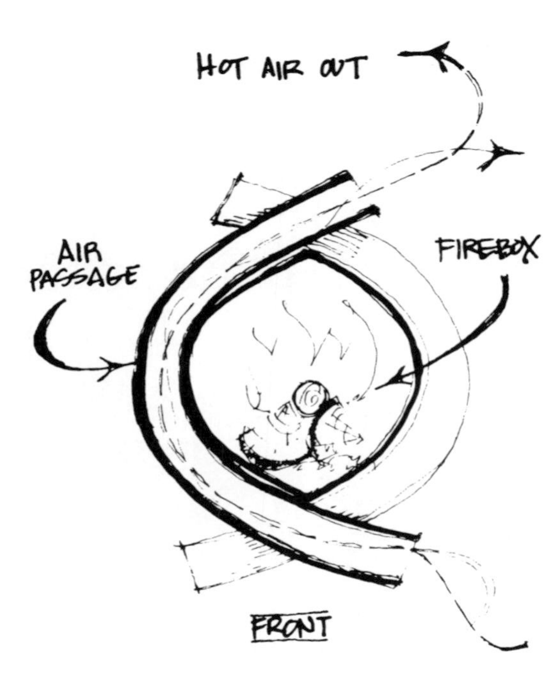

one pound of water one degree Fahrenheit. An average house loses about 20,000 BTU per hour on a typical winter day.

Calorimetry. Direct measurement of the heat output of a stove, usually through the use of a calibrated and instrumented room. In order to calculate overall efficiency, one compares the measured heat output of a stove with the known heat potential of the fuel consumed by the stove.

Catalytic Combustor. A device employing a catalyst that can promote ignition of air/exhaust gas mixtures at temperatures lower than normally would be required.

Charcoaling. The process of forming charcoal by driving off volatile material through devolatization and/or pyrolysis.

Convection Stove. A stove which transfers the majority of its heat through convection. This is usually accomplished by designing an air passage between the hot firebox and a decorative shell or shroud. Cool room air can then travel through this passage by natural or forced convection, picking up heat and carrying it into the room.

Cross-draft Stove. A stove design in which combustion air and exhaust gases enter and leave the firebox at approximately the same height (usually in the lower portion of the stove), and result in a horizontal burn path.

Devolatization. The process occurring when wood or highly volatile coal is gradually heated to approximately 500°F–600°F, releasing volatile gases and leaving charcoal or coke.

Down-draft Stove. A stove in which combustion air enters the firebox above the grate and is drawn down through the fire-mass, exiting below the grate.

Draft Inducer. A mechanical device that is used to create a draft where a natural draft would not occur, or where a natural draft is insufficient.

Efficiency. A measure of a stove's ability to utilize a fuel's full energy potential. *Overall Efficiency* is the ratio between the heat which actually enters the room and the total heat available from the fuel. It can be broken into two components, *Combustion Efficiency* and *Heat Transfer Efficiency*. Combustion efficiency is a measure of how completely the fuel has burned. Heat transfer efficiency is a measure of how much of the heat produced is transferred into the room.

Overall Efficiency =

Combustion Efficiency × Heat Transfer Efficiency (See Emissions.)

Emissions. All substances discharged into the air during combustion. These substances can be broken down into two categories: *Gaseous Emissions* and *Particulate Emissions.* Typical gaseous emissions include carbon dioxide, carbon monoxide, water vapor, and hydrocarbons. Typical particulate emissions include fly ash, tar aerosols, carbonaceous soot, and polycyclic organic molecules (POM's), some of which have been implicated as carcinogens. All emissions except carbon dioxide, water, and fly ash represent combustion efficiency losses.

Excess Air. Air (oxygen) that is not used for combustion entering a stove. This extra air cools the firebox, lowers combustion efficiency, and carries heat up the flue to lower heat transfer efficiency. A small amount of excess air is desirable as a means of insuring that adequate oxygen is available for complete combustion.

Firebox. The area within the heating appliance used to contain the burning fuel.

Firebrick. A brick made of materials capable of withstanding high temperatures for extended periods of time. It is used to build fireplaces and to line furnaces and wood stoves. (See Refractory.)

Fireplace Insert. A solid fuel combustion device that is placed in a fireplace cavity. It offers improved efficiency over the open fireplace by reducing excess air through an air control device similar to a woodstove. The insert can also improve the heat transfer of the fireplace, usually through the use of a blower that forces air around the firebox. Most units still offer an unobstructed view of the fire.

Gasification. The process of converting a solid fuel such as wood or coal into a combustible gas.

Grate. A frame, bed, or platform within a stove on which a fire is made.

Heat Exchanger. Any device designed to transfer heat. An area of (or within) a stove where hot exhaust gases can transfer their heat to room air or to incoming combustion air.

Heat Transfer or Exchange. The process whereby heat moves from hot areas or objects to cooler areas or objects. There are three types of heat exchange: conduction, convection, and radiation. *Conduction* is the transfer of heat through a material without any appreciable movement of its molecules. Heat is transferred through a stove wall by conduction. *Convection* is the transfer of heat by the moving and mixing of a fluid (air).

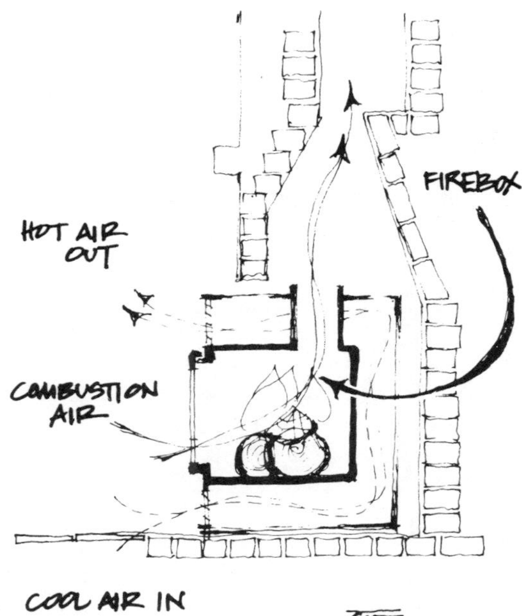

Fireplace Insert

This movement can be caused by two means: *natural convection* and *forced convection*. Natural convection occurs when hotter, less dense air rises and is replaced by cooler air in a continuous cycle, such as what occurs next to the body of a hot wood stove. Forced convection employs fans or blowers to move the air. *Radiation* is the transfer of heat in the form of electromagnetic waves (infrared) similar to light waves and involving no mechanical means of transfer. Not until the radiation is absorbed by an object is the heat noticeable.

Heating Value. The amount of heat that has evolved per unit of fuel when the fuel has completely combusted. This value is usually expressed for solid fuels in BTU's per pound. The *higher heating value* of a fuel indicates that all water formed during combustion condenses into liquid form. The *lower heating value* indicates that water of combustion remains as a vapor. For combustion of a fuel such as wood, the heating value must take into account the energy lost due to evaporation of inherent moisture. For example, one pound of perfectly dry wood contains 8,600 BTU's; one pound of 40% moisture content wood (wet basis) contains only 4,760 BTU's of usable heat or 55% of the heating potential of the dry wood.

Ignition Temperature. The temperature at which a substance will spontaneously ignite in the presence of sufficient oxygen.

Kilocalories (1000 calories). A quantitative measure of heat where one calorie is equal to the amount of heat needed to raise one gram of water one degree Celcius. One kilocalorie equals 3,968 BTU's.

Kilowatt (1000 watts). A measure of power output also used to indicate heat. One kilowatt equals 3,415 BTU's.

Products of Combustion. Substances formed during combustion of a fuel. The products of complete combustion are carbon dioxide and water. Products of incomplete combustion can include carbon monoxide, hydrocarbons, soot, tars, etc.

Preheated Combustion Air. Air which passes through some type of heat exchanger prior to being involved in combustion; e.g., preheated secondary air.

Primary Air. Air entering the stove, used to maintain combustion in the firebox.

Radiant Stove. A stove that emits most of its heat in the form of infrared radiation.

Refractory. A ceramic material used to line stoves and furnaces to protect metal parts from extreme heat and oxidation. A refractory may also be formulated with insulating qualities in order to maintain maximum temperatures in certain areas of the stove for more complete combustion, as in the case of refractory-lined combustion chamber.

Scandinavian-type Stove (Cigar burn; front to back). A front-loading box stove employing a horizontal baffle located in the upper portion of the stove. This baffle forces gases to exit via an opening in the upper front of the firebox. This design tends to encourage a burning of the wood load from front to back similar to a cigar.

Secondary Chamber. An area within a stove, usually adjacent to the firebox, that is used for the combustion of exhaust gases.

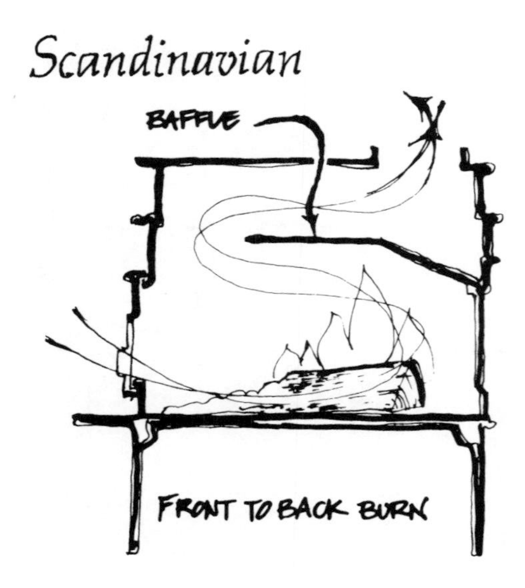

Secondary Combustion. Combustion of unburned gases emitted from the primary firemass, usually in a secondary chamber. Under normal circumstances a temperature of at least 1100°F must be maintained along with sufficient quantities of oxygen, well mixed with the exhaust, to accomplish secondary combustion.

Stack Loss. Heat and potential heat that is lost up the stack.

Step. A front-loading box stove employing a "stepped" top plate.

Thermal Conductivity. The measure of a material's ability to conduct heat. Metal has a high thermal conductivity and is a good medium for transferring heat. Insulating refractories have a low thermal conductivity and are good for retaining heat within an area.

Thermocouple. A temperature-sensing device employing two wires of a dissimilar metal welded at one end. When the junction of the metals is heated, it produces a known millivoltage output that can be measured and used to determine temperature. Typically, a wire probe, a reference junction, and a meter are used.

Thermostat. A device used to control heat output from a heating appliance. Typically in a woodstove, it accomplishes this by responding to temperature changes with movement—by utilizing a bimetallic coil or strip (two metals with different expansion rates joined together). The movement is then used to operate some type of combustion air control device such as a flapper.

Up-draft. A stove in which air enters relatively low in the firebox or under the grate and travels around or through the fire-mass before exiting near the top of the stove.

Volatiles. Gaseous and liquid materials that are driven from wood (or certain coals) as they are heated to approximately 500–600°F. Wood volatiles usually contain various hydrocarbons and can represent up to one third or more of the wood's energy.

Up-Draft

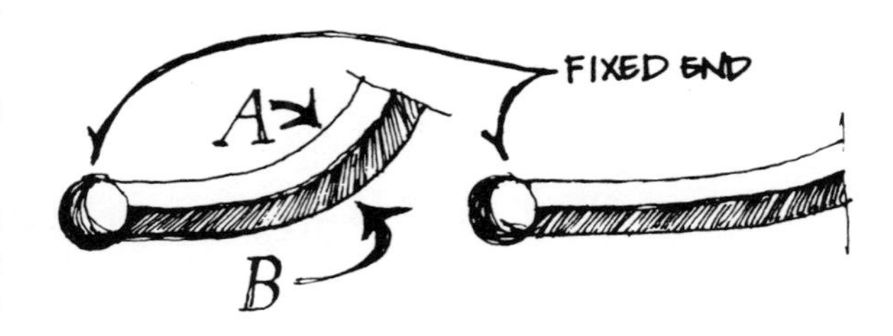

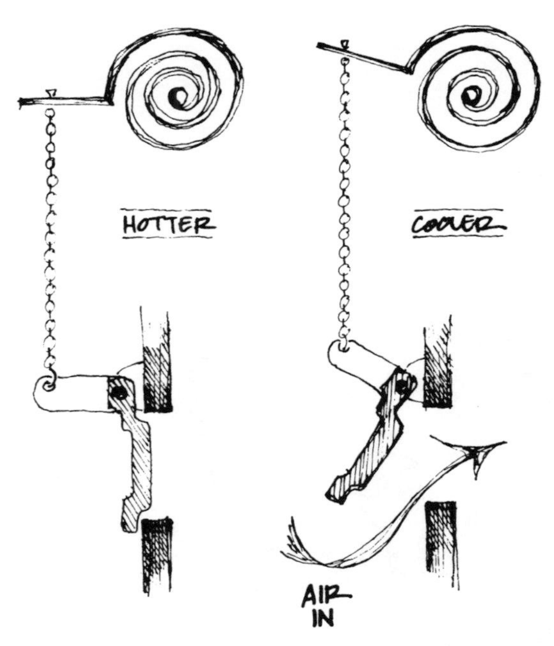

Bimetallic Strip Thermostat

Independent Study in Maintenance

Every Fall the Randolph Fire Department has a series of dress rehearsals for the coming winter when they are called out to extinguish chimney fires. The fires are invariably the result of poor planning on the part of a local woodburner, who should have cleaned the flue thoroughly the previous spring but was too busy working up wood or putting in the garden. Suddenly the cool nights of October arrive, and the first hot fire touches off last spring's accumulation of soot in the chimney.

We place the system maintenance section under Fall in this book as an emergency alert rather than a recommendation. If you haven't thoroughly cleaned the system yet, *do it now!* Hopefully the cleaning job will be done in the spring each year. It is a compromise to procrastinate the job until fall, and it is downright unsafe to enter a heating season without having cleaned the flue from the previous year.

The Chimney

How often should the chimney be cleaned? This is one of the first questions the safety-conscious woodburners will ask. The best answer is "when it needs it."

Traditional advice was that the chimney should be brushed once a year. This standard, however, was developed when most people were burning wood in fireplaces or non-airtight stoves that lost large amounts of heat up the flue; this prevented creosote from forming. Modern airtight stoves are more efficient, and this means less heat lost up the flue and greater levels of flue gas condensation. The result is the need for more regular cleaning. (Coalburners do not need to worry about creosote formation, but should clean the chimney at least once a year to prevent any corrosive action.)

The frequency of cleaning will vary, depending on many factors. Two stoves in the same house may require different cleaning schedules depending on the stove type, the chimney, and the operating techniques. Most experienced woodburners constantly monitor their installation for creosote buildup. This can be done best visually with a mirror and flashlight, but you can also learn to diagnose the condition of your pipe by tapping on it with a fingernail or ring. A clean pipe gives a metallic ring; a pipe in need of cleaning offers a muffled thud. Knock sharply on the pipe with your knuckles; if you hear what sounds like potato chips falling you will want to clean it soon.

Creosote can build up at an impressive rate. A few hours of a cool flue and a smoldering fire can produce more black liquid than you thought possible; a few days of low-level burning can form a web of creosote that may completely seal the flue.

Once you have determined that the chimney needs cleaning, you will have to decide who is going to do it: a professional chimney sweep or yourself. Hiring a sweep will make the most sense for many people. Sweeps know how to get the job done right and as a group are conscientious, independent businessmen who will place a priority on your satisfaction with their work. If you are unable to or prefer not to work at heights, if your chimney has difficult access, or you object to doing a completely dirty job, call a sweep. If none of the above characteristics describes your situation, then do the work yourself. You will truly be able to do the cleaning "when it needs it" and will save money in the process.

There are a number of ways to clean a chimney. One approach that may be more myth than method involves tying a piece of clothesline to a live chicken. The unfortunate bird is inserted into the flue and its flapping wings dislodge the soot. It is claimed that you can get two flues to the chicken plus the evening meal.

You can also clean the flue by pulling through it a large bag filled with rocks, straw or twigs, or you can rattle a tire chain fastened to a rope throughout the cavity. The list of alternative methods continues with regional claims that great success can be had by throwing such things as flashlight batteries or potato skins into the burning stove, or by using commercial chimney cleaning chemicals. One country method that is perhaps the most dramatic is to start a chimney fire on purpose once a year. While this last method will no doubt do a thorough cleaning, it presents the risk of setting the house as well as the soot on fire.

None of these old-time procedures is worth pursuing since good quality chimney-cleaning equipment has become available in recent years. The equipment selection includes individually sized brushes to fit your flue and flexible fiberglass rods that can be inserted at angles through the thimble. The entire selection of equipment you will need will pay for itself in savings from a professional service in the first year. A coopera-

Chimney Brushes

Chimney Cleaning Equipment

Hiring professional sweeps to clean your chimneys twice a year can be an expensive proposition. This rugged, high-quality equipment allows you to do the job yourself and should pay for itself in one heating season.

The brushes have the scouring power needed to remove stubborn creosote and soot. The fiberglass rods are heavier than most on the market, yet are extremely flexible.

We offer two types of brushes. For wood burners, we have a "round-wire" brush. The bristles are made of heavy-gauge, uniform-length wire for maximum cleaning. For the coal burner, a stiff, black polypropylene brush provides the wiping action required to remove fine particles of soot. Unlike a wire brush, the polypropylene is impervious to the corrosive elements of coal soot.

How to Order

It's vital that you accurately measure the size of your chimney. To determine the correct brush size, measure the interior of the flue to the nearest ¼-inch. Almost all round flues will measure to a full inch dimension. Rectangular, tile-lined flues, however, will usually not be a full-inch size and require careful measuring. The rule of thumb in the case of odd size flues is as follows: if your flue is less than ¾" to the next full inch size, select a smaller brush. If your flue measurement is ¾" or above, select the next full-inch brush size.

Example: A rectangular chimney measures 6½" x 7¾". Choose a 6" x 8" brush. The flue is too narrow to accept a 7" brush, but will take an 8" brush in the 7¾" dimension.

Example: A rectangular flue measures 8½" x 8". Select an 8" x 8" brush.

Example: A round flue measures 8" in diameter. Select an 8" round brush.

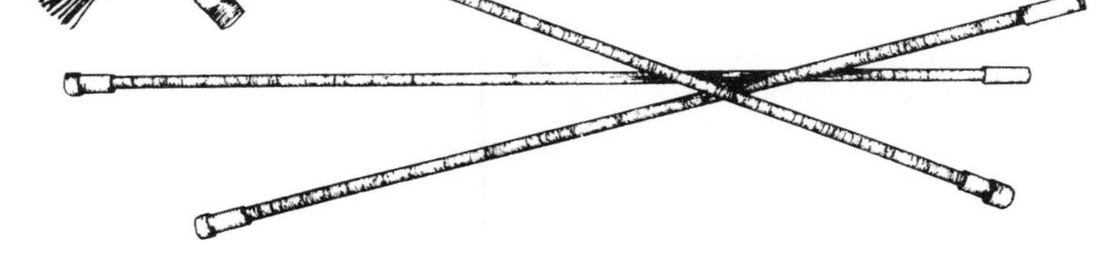

tive purchase with one or more neighbors will allow you to save even more.

The creosote in your flue may be either a light, crispy flake composition or a thick, baked-on tar. The former will be dislodged readily with your brush; the latter will require considerable scraping by hand. You may want to call on the expertise of a chimney sweep if deposits remain in the flue despite your best efforts.

Creosote stains are difficult to remove from the external surface of an installation. If well hardened, the material may be scraped or sanded off with good results. There does not yet seem to be developed a cleaning agent that will do a complete job of removing stubborn creosote stains.

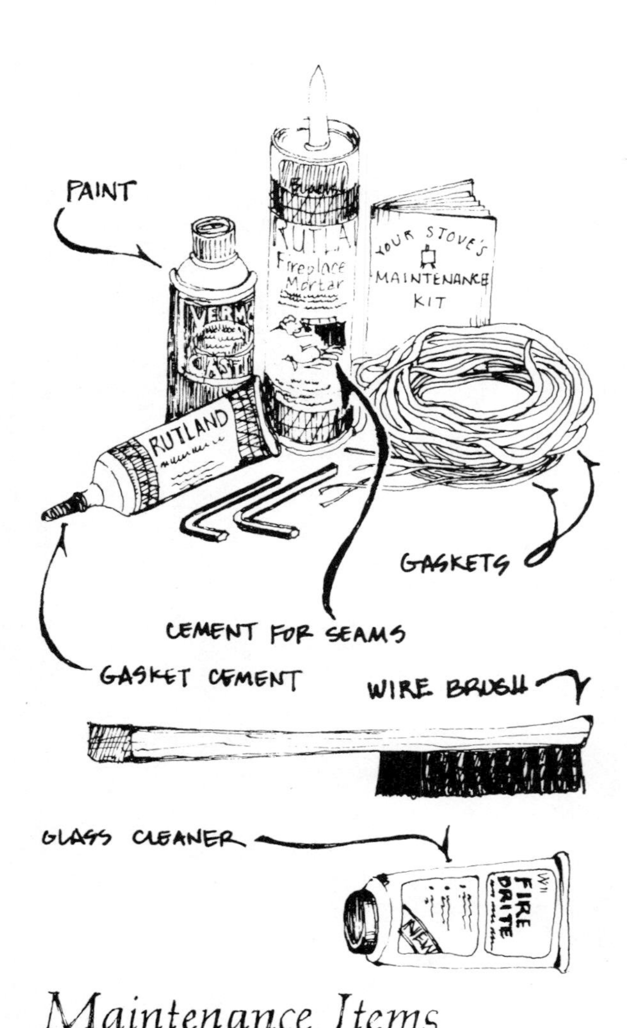

Maintenance Items

The Stove

Wood and coal stoves are generally simple machines that can be serviced with an hour's time and effort. Clean your stove by first removing the ashes; make sure you leave enough for a good insulating layer. Brush down the entire inside of the stove. If the stove has been burned hot, the deposits on the inside of the stove should be a light fly ash that are easily removed. If the interior has a buildup of creosote, this task requires a stiff wire brush and a little "elbow grease." Inspect the interior for any damage that has occurred from last winter's use. Any part that comes in contact with a hot wood or coal fire should be considered a wear part, and you should expect to replace these parts on a periodic basis. The schedule for this replacement will depend on how hard your stove is used. If you use your stove as the sole source of heat in a northern climate, the parts will obviously need to be replaced more often than a stove that is used occasionally as supplementary heat in a mild area.

Next, check the condition of the gaskets. As you look at the inside of the doors and griddles, look for white areas on the plates. These white areas may indicate gasket leaks. If the gaskets are dark and brittle they should be replaced. Fiberglass gaskets should be replaced annually. Asbestos rope gasketing usually has a longer life and should be replaced as needed. When gasket leaks are discovered, they may be shimmed in the appropriate area with thin strips of gasketing inserted under the existing gasket. If this fails to stop the leak, the complete regasketing of the door or griddle may be indicated. Most doors have a mechanism to adjust the tightness of the seal. Proper door adjustment may solve most leaking problems. Check the moving parts of the stove. The air controls should move freely. High temperature silica grease may be applied if needed. If your stove has a thermostat, it should be inspected to insure that the lever moves and that the coil will still react to heat. This can be checked with a match.

Check for missing furnace cement on stoves that use it as a seal. In time, furnace cement may harden and crack out of the joints of many stoves. This can be repaired easily with a "fingerful" of furnace cement forced into the seam. Be sure to use a water soluble cement so you will be able to wash the excess off the stove with a wet rag.

You are now ready for the final step of painting the stove. Many people make the mistake of using stove black or stove polish on a stove that was originally painted. If you choose to do this, you should realize that since the polish has a petroleum base you will not be able to paint over it. Polish will collect dust and fly ash more easily and will rub off the stove with use. Instead, touch up or repaint the entire stove with a high temperature spray paint. This process takes about ten minutes, and the stove will once again look like new.

Remove rust stains from a cast iron griddle with a light wire brushing and treat it with vegetable oil or suet. Finish up by polishing the handles and trim with metal polish, and the job is done. You are now ready for winter. Next year, remember to do all of your maintenance in late spring or as soon as the heating season is over.

A Coal Primer

Students of solid fuel heating who have completed the "Coal Burning for Beginners" section will want to broaden their base of understanding of coal and how it burns. The sensation of coal heat is slightly different from wood: wood burns in combustion cycles that feature peaks and valleys of heat. The fire starts from a low point and is built up until the stove is hot. The fuel burns for several hours, and the temperature begins to decline. More fuel is added, and the heat level climbs again.

A coal fire, on the other hand, begins very gradually. Once the fire is well established and the stove is hot, however, the heat level remains remarkably consistent. The coal stove that cranks out heat hour after hour will impress even the skeptical stoveside observer.

The Past

Anyone whose memory includes a recollection of life in the first half of the twentieth century may be justifiably amused at the current notion that coal is a "new" energy resource. It is anything but new. Oil, gas, and electricity have been popular forms of home heating only for the last thirty years or so. Before then, burning coal (or wood) was how you stayed warm during the winter.

Although the availability of wood has made it the primary choice for fuel from the time when our primitive ancestors first discovered fire, coal has been popular for at least 500 years. Traditionally, coal has been the fuel that man has turned to when wood supplies became depleted. This was observed first in England in the sixteenth century. A general population drift to the city depleted the fuel resource of the surrounding forests. At the same time, the roads were deteriorating from a series of especially severe winters. Since wood was not immediately available and transporting it by road was not feasible, people turned to coal that could be shipped by small coastal barges from the north.

About the same time, another development made the use of coal more attractive. Coal fires then, as now, required adequate chimneys and a good draft. Since chimneys at that time were made out of iron they were expensive, and only the wealthy could afford them. Around 1600 the process of making bricks in coal-fired lime kilns was perfected. This new supply of inexpensive bricks made it possible for nearly everyone to have a chimney that allowed them to heat and cook with coal. Coal remained a popular fuel in the urban areas of Europe until after 1950.

The American Indians knew about coal long before it was "discovered" in Illinois in the 1680's. Bituminous was first mined in Virginia in 1745 while anthracite was mined in Pennsylvania by 1790. It wasn't until the mid 1800's that coal became a popular alternative to wood. As in England, this conversion was stimulated by the growing population of the cities and the subsequent diminishment of the wood supply.

The same trend repeated itself in the 1970's. As the cost of oil, gas, and electricity soared to record highs, people again began burning wood. By the 1980's, the ranks of woodburners had swollen to such levels that in some urban and suburban areas the cost of wood rose also, and the supply became uncertain. As in the past, coal came to the fore as the fuel most readily available and economical to use.

Coal Sizes

After it is mined, coal is crushed and screened into various sizes for industrial use, power generation and home heating needs. While power generating plants might buy a large sized coal and crush or pulverize it to the correct size for the furnaces, most coal destined for industrial and dometic use is crushed at a processing station near the mines, then sorted into standard sizes.

The size of the coal burned will influence the stove's heat output and responsiveness. Generally, the larger sizes of anthracite produce more heat while smaller pieces are easier to light and will rekindle faster after a fresh load of coal is added to the fire.

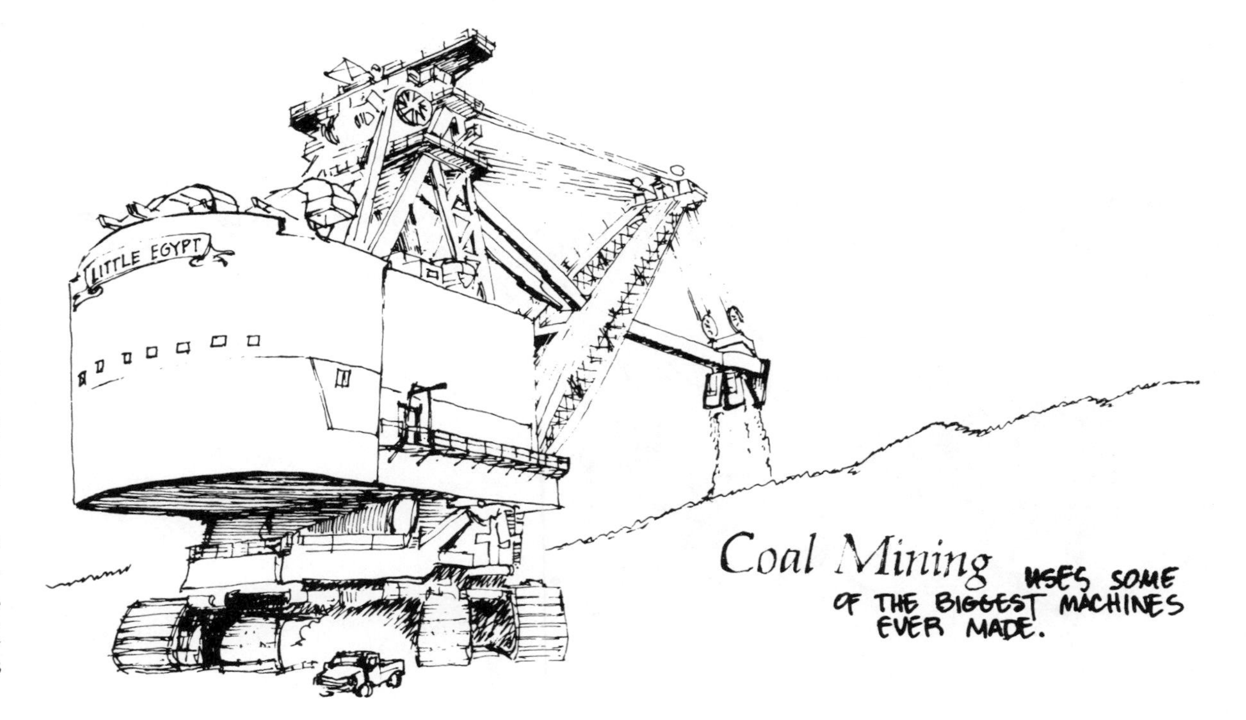

Pea, nut, and stove sizes of anthracite are the most widely used sizes for coal stoves while smaller sizes such as buckwheat, rice, and barley are used for auger-feed stoker systems that are more common in large scale applications.

Undersized or oversized coal can be hard to deal with. According to anthracite sizing standards 15% of the coal leaving the breaker, where it is crushed into various standard sizes, can be undersized while about 10% can be oversized. In practice, coal sizes can vary a great deal more than this.

Coal that is smaller than required or mixed sizes can make the fuel bed more compact. Since less air can pass through for combustion, the fire will tend to burn slowly and react slowly to changes in the air supply. Sprinkling small pieces of coal and coal dust, referred to as "fines," over the coal fire at night is an old trick used to bank the fire. Too many fines dispersed within the coal has the same effect of slowing the rate of burn. Larger than required coal, on the other hand, creates large air spaces between the coal that can make it burn too quickly. Excessively large or small coal sizes may be mixed together to achieve more consistent performance.

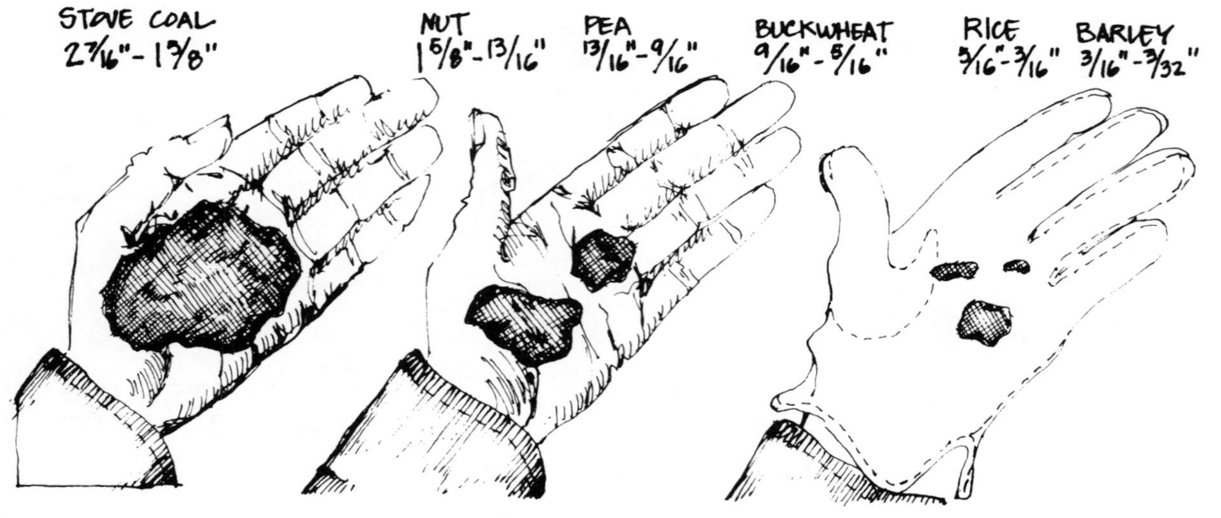

Coal Type and Quality

Unlike wood, which tends to have predictable burning characteristics, the combustion of coal can vary significantly depending on the type and quality used. The composition differs because coal is formed from varying organic and inorganic materials under a wide range of pressure and temperature conditions. Coal quality and content can differ depending on the geographic area, the individual coal field, the specific coal seam, and may even differ within the same seam.

In order to differentiate between coals with similar characteristics, the American Society for Testing and Materials (ASTM) has developed a ranking system. In this system, coal classification is based on progressive degrees of difference from soft coal (lignite and bituminous coal) to hard coal (anthracite coal). Coals that have the most fixed carbon, or carbon that stays in place as incandescent coal during the burning process, are classified by the amount of fixed carbon they contain on a dry basis. (Moisture trapped within the coal is not included in the analysis.) Anthracite coals and higher-ranking bituminous coals are classified in this fashion. All other coals are classified according to their calorific value on a moist basis. (That is, inherent moisture that is trapped within the coal is included in the analysis and is considered in determining the coal's BTU value.) The noncombustible material in the coal, such as ash, is referred to as mineral matter and is not included in the determination of rank.

Chemical Composition

As you move from one coal grade to the next, there is a gradual change in the percentages of volatile matter, fixed carbon, and moisture. (Volatile matter is the part of the coal that is liberated as a gas during combustion.) These three constituents of coal are the criteria included in a proximate analysis, and as the percentage of each changes, so does the way the coal will burn.

Anthracite coals, for instance, are high in fixed carbon content (above 86%) and low in both volatile content (usually below 10%) and in moisture content (usually below 3%). The coal burns with a short blue flame, indicating that the small amount of combustible gas that is released and burned above the fire is carbon monoxide. The heating value is high due to the high percentage of fixed carbon and the low moisture content. As you move into the softer coals, the percentage of fixed carbon decreases and the percentage of volatile matter and inherent moisture increases. While volatile matter for anthracite generally will be below 10%, most bituminous coals range from about 16% to about 40% volatile content. When bituminous coals burn in a typical coal stove or furnace, long yellow flames are produced, indicating that most of the combustible gases are complex hydrocarbons that are not burning completely over the fire. Sub-bituminous and lignite, which are considered low-ranking coals, burn with similar long yellow flames, yet their heating value is lower due to a high moisture content. (Some lignite coals have 25% or more moisture content.)

Cannel coal, which is not included in the ranking system, is a very high volatile (50%), hard, bituminous-like coal that is used primarily for open fireplace use. Due to its high volatile content, it is not recommended for use in stoves and furnaces.

When coal is sold wholesale, either at the breaker or from larger coal suppliers, a proximate analysis is conducted by an independent laboratory and this is provided to the buyer. If you are buying a winter's supply of coal at one time, it would be helpful if you obtained the analysis from the coal dealer—knowledgeable dealers will know what this is, or they will be able to tell you what the ranges are for fixed carbon, volatile matter, and moisture in the coal they are selling.

You can learn a lot about how a load of coal will burn from this analysis. The volatile content, for instance, affects how easily the coal fire lights and how responsive the fire will be when fresh coal is added to it. Conversely, the higher the percentage of fixed carbon, the more difficult the fire will be to light. This is particularly true for anthracite.

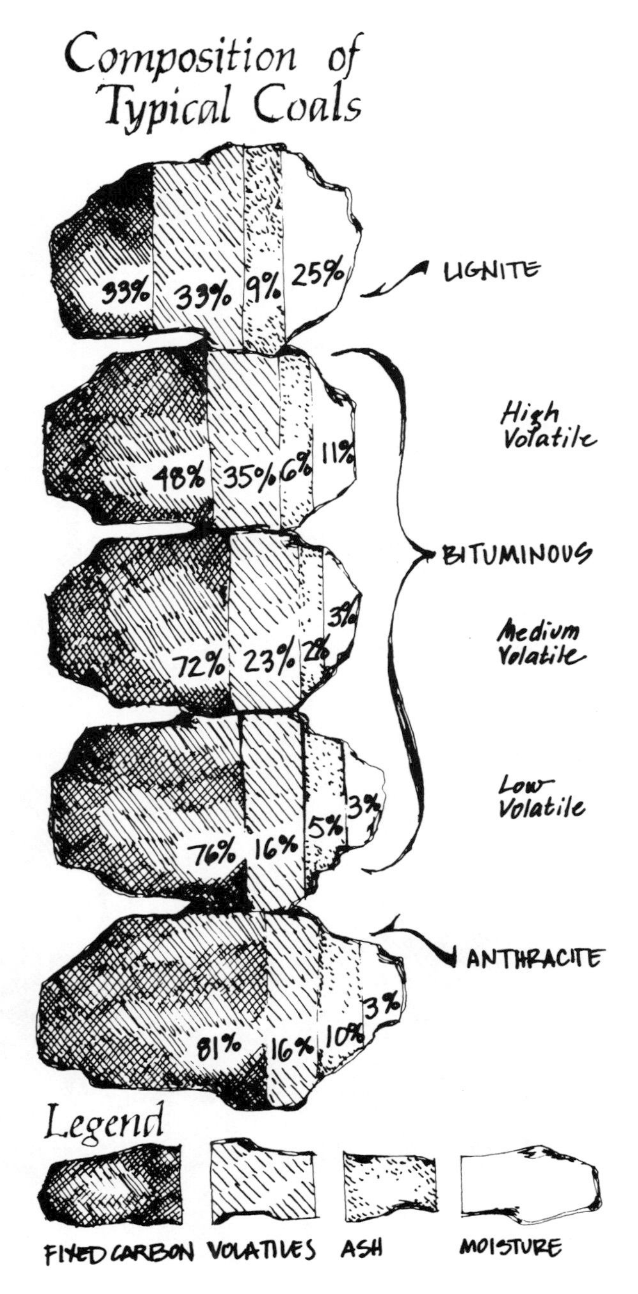

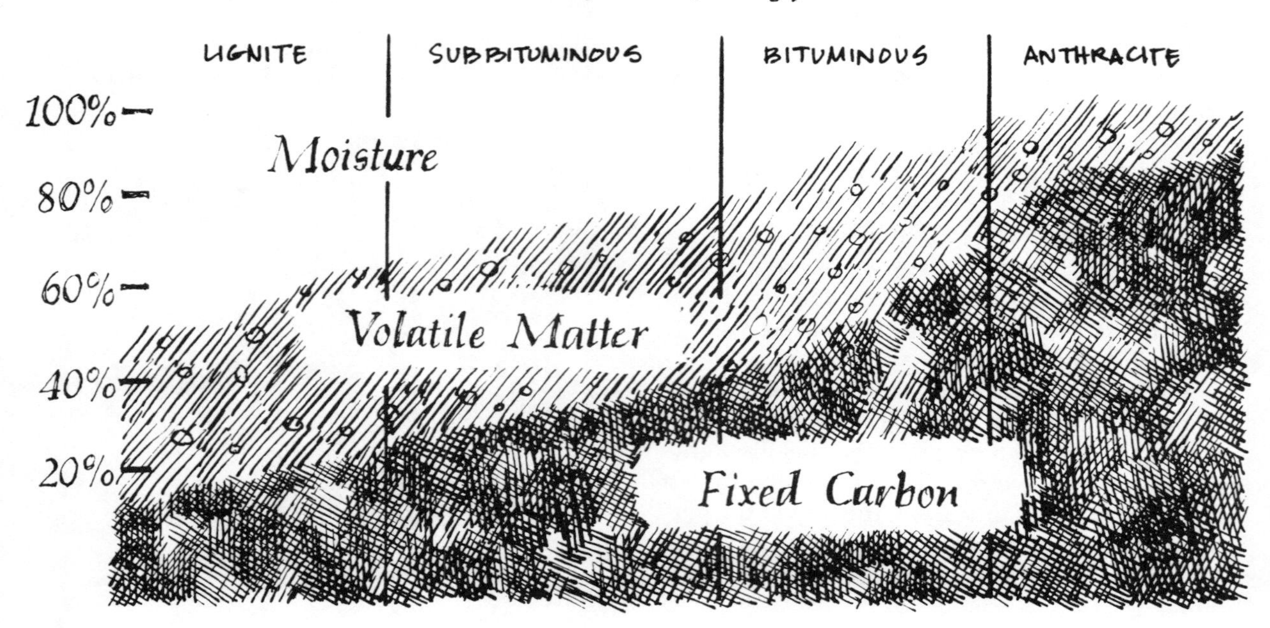

In the typical range for commonly available anthracite of between 5% and 10% volatile content, the closer you are to 10%, generally, the easier the coal is to light and to keep lit. Due to its low volatile content, anthracite is considered the cleanest burning coal, and you rarely see smoke coming from chimneys where anthracite is being burned.

For bituminous, sub-bituminous and lignite coals, the higher the percentage of volatile matter, the more hydrocarbons are released during the burning process. As the coal is heated, these volatiles, in the form of tar vapors and soot particles, are driven off in the same manner as the volatiles are given off from wood.

It is the relatively high volatile content of bituminous, sub-bituminous, and lignite coals that make them dirty, and in some cases unsafe, to burn in many stoves and furnaces presently available in the United States. When these grades are added to a hot bed of coals, it is the volatiles that are released first in the combustion process. Although there may be air available to burn these gases, the fresh load of coal has reduced the temperatures above the fuel bed to the point where complete combustion is impossible. The deeper the layer of fresh coal, the greater the length of time these volatiles will be released during the burn cycle. Even a thin layer of fresh coal will produce volatiles that are not burned.

The results of burning bituminous coals in simple stoves and furnaces that are not specifically designed to burn these volatiles can be disastrous. A build-up of volatile gases over the fire can cause explosions within the appliance similar to backpuffing with wood stoves.

A more serious problem, though, lies in the large volume of soot and hydrocarbons that pass through the flue system and into the atmosphere. Deposits in flue pipes and in chimneys, which can actually block the flue in as little as two weeks, are extremely dangerous. The soot deposits can also ignite and cause chimney fires. The smoke that is produced can be significantly worse as an air pollutant than either smoke from anthracite coal or from wood.

After the volatiles are released, the remaining coal, which is mostly fixed carbon, burns clearly until all that is left is ash.

The moisture content shown in the proximate analysis represents the moisture that is trapped within the coal. Anthracite coals have very small amounts of moisture while sub-bituminous and lignite coals can have significantly higher amounts. This water is driven off during the combustion process and lowers the heating value of the coal. This is the major reason why it is not economical to transport lignite and sub-bituminous coals to distant power plants; for a lignite coal with 25% moisture content, 25 out of 100 carloads of coal are, in effect, water.

There are other factors that are not listed in a proximate analysis that influence the suitability of a coal for home use. The coal's ash content and the nature of the ash affect the frequency with which the fire must be tended to keep it going. While wood might have an ash content of 1%, the ash content of commonly available coal can range from about 5% to 15%. The higher the ash content of the coal, the more often the ashes will have to be removed from the stove or furnace firebed to maintain a given heat output.

The highest quality anthracite will have an ash content below 10%. Medium quality anthracite ranges between 10% and 13%, and poor quality over 13%. The ash content of commonly available bituminous coals is usually in the 5% to 10% range. Burning coal with a high ash content will cause problems when kindling and maintaining the fire.

Poor quality anthracite coal has an ash that tends to stay in place after some of the fixed carbon has been burned away. This coarse ash prevents air from reaching all the fixed carbon at the center of the individual coal piece, which in turn lowers combustion efficiency. Also, this type of ash is more difficult to remove from the fuel bed. The resulting "dead" zone, where air coming through the fuel bed has been blocked, makes it much more difficult to keep the fire going.

FOR A GIVEN AMOUNT OF HEAT

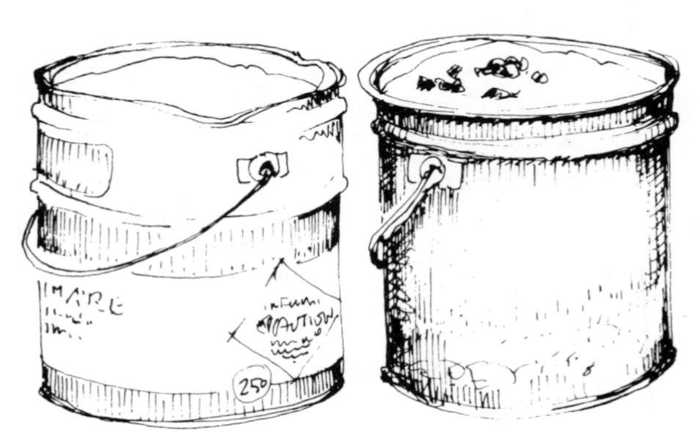

COAL PRODUCES TWICE AS MUCH ASH BY VOLUME AS DOES WOOD

Managing the Two Acre Coal Mine

The convenience of handling coal may in one way deprive the coal burner of a benefit enjoyed by his woodburning counterpart: a plethora of literature on harvesting, handling, cutting, and splitting the stuff. Let's suppose, though, that one could get as intimately involved with procuring the coal supply as with wood. It might sound something like this:

Digging your own coal can be a rewarding and satisfying project if you use the correct tools, are safety-conscious, and enjoy self-sufficiency. The dark confines of your private shaft can become something of a refuge from the trials of everyday life, and can provide you with free fuel as well.

Coal mining is not for everyone, however; it will depend on your geographic area and the availability of coal reserves under your state. A Vermonter, for example would have to dig a vertical tunnel a considerable distance to the south before he could expect to strike a seam. Check with your state geologist about coal seams in your town before you start to dig.

The small-scale miner will want to start by investing in the proper tools. A pick axe is the traditional mining implement; get one as heavy as you can lift, with a hardwood handle and a good sharp point. You will also need a sturdy shovel; select the type that has a steel scoop, with the sides of the scoop flared upward to prevent the nuggets from spilling over the sides. Don't try to compromise on this item by trying to use your snow shovel; it just won't hold up. Add to this an eight-pound sledge hammer to reduce the larger chunks to stove size and you're almost ready to dig.

Next comes proper clothing. A hard hat is indispensable. Not only will it protect your head from bumping the top of the shaft as you stand up, but it will protect you from poorly planned swings of the pick axe. A headlamp is also essential. If the helmet does not come equipped with one, buy the type favored by nocturnal skiers that straps to your head with an elastic band and operates on batteries. A pair of steel-toed boots and work gloves will round out your subterranean wardrobe.

You will need some sort of container to remove the coal from the mine. A couple of sturdy coal hods will do for starters, but a wheelbarrow will make the work easier. Should you find that your neighbors are asking for a ton or two each year, the traditional mule and cart may be something to consider.

Always wear a protective respiratory mask and never enter the mine without some method of detecting methane gas. A small gas-detecting meter is handy but a canary will do. The latter will keep you company as you work, but the meter has the advantage of being reusable.

Laying in the coal supply requires discipline and persistence. The average home will use, say four tons of coal each year for heat. This is 8,000 pounds of coal. A miner of average strength will be able to carry a full coal hod out with each trip, for a total of 100 pounds per trip. This is 80 trips, which means that on some weekends you will have to double up your efforts.

As the price of other fuels continues to rise and availability remains uncertain, you will easily find a market for extra coal you may have. Once you have established a reputation as a producer, however, you should be prepared to cope with government regulations and possible pressure to unionize yourself.

Most weekend coal miners do it just for the fun, though . . . and the reassurance that they will continue to be warm if the supply lines break down.

Coal ash melts and forms "clinkers" when the temperature within the fuel bed exceeds the ash fusion temperature of the coal. The ash fusion temperature depends on a number of factors relating to the mineral content of the ash and the amount of oxygen present in the combustion zone. Coal with an ash fusion temperature above 2700°F is considered to be the best; 2200°F to 2700°F is medium; and below 2200°F is less desirable. With good quality anthracite, coal clinkers rarely form; when they do, it's because the stove or furnace was overfired.

Some bituminous coals devolatize and then fuse together when they are heated. This process is called agglomeration and is an undesirable characteristic because the coal then blocks incoming air and creates "blow holes" in the fuel bed, resulting in incomplete combustion. Coals that do not agglomerate are more desirable and are referred to as "non-caking" or free-burning coals.

Another characteristic of coal that affects how easy it is to handle has to do with how soft or hard the coal pieces are. "Friable" coals break up easily during handling and transporting. A highly friable coal could be shipped as nut sized coal and be delivered as a fine powder because of its weak structure. Generally, low volatile bituminous coals fall into this category; however, some of the higher volatile anthracites (10% volatile) can have a weak enough structure to form excess coal dust with handling.

The percentage of sulfur that is present in the coal determines the amount of sulfur oxide pollutants generated as well as the amount of sulfur available to form sulfuric acid within the flue system. Obviously the lower the sulfur content in this respect, the better the coal is. When sulfur is burned, sulfur dioxide and sulfur trioxide are formed. When the dew point of these gases is reached, (called the "acid dew point"), they will condense on flue pipes and on chimney walls. With the addition of moisture from the combustion process and from humidity in the air, they form sulfuric acid. The acid is trapped by soot and fly ash deposits within the flue system to form what is referred to as an "acid smut" that can corrode some prefabricated metal chimneys and even masonry chimneys over a period of time. Aside from having your chimney cleaned periodically (at least once each spring), choose a coal with a low sulfur content to minimize this problem. Generally, anthracite coal has a sulfur content below .5% while some eastern bituminous coals have sulfur contents in the 2.5% to 4% range. Western coals generally have a sulfur content below 1%.

Your coal dealer should be able to tell you about the composition of his coal. However, the best way to determine how well the coal will burn in your stove or furnace is to burn it. Purchase two or three hundred pounds initially; then, if the coal burns well, purchase the rest. (It's no fun shoveling out a coal bin of unsuitable or poor quality coal.)

Stoves and Furnaces

Historically, stoves and furnaces were designed for specific types and qualities of coal. There are a variety of basic designs that are now being used throughout the United States.

Fireplace Coal Baskets. Coal baskets are designed for cannel coal (a high volatile, bituminous-like coal), bituminous and lignite, free burning coals (non-agglomerating), and for wood. Since combustion air is not directed only through the basket, baskets cannot be used for low volatile fuels such as anthracite coal and coke (devolatized bituminous coal).

Surface-fired Stoves and Furnaces. These units are commonly referred to as "chunk burners" or "hand fired" units. Coal is added to the combustion zone through a loading door directly above the fire. Primary air is supplied under the bottom grates, and the air needed to burn volatiles is added above the fire. Although this system has been used to burn virtually all sizes and types of coal, there is no effective way to regulate either the secondary air above the fire or the depth of the fuel bed, making it the worst potential offender from an air pollution standpoint.

Gravity-feed Surface Fired Appliances. These stoves and furnaces are designed primarily for anthracite coals. Fresh coal is loaded into the combustion zone in thin layers. The main advantage of gravity feed systems is that they maintain the proper fuel bed depth for efficient combustion, and by having a regulated bed depth for each size coal, the air flow through the fire can be precisely regulated. A disadvantage of gravity feed stoves is that they require a specific size of coal, generally nut or pea size.

Horizontal and Down-draft Combustion Appliances. In these stoves and furnaces the incoming air passes horizontally through the fire and into a refractory-lined secondary combustion

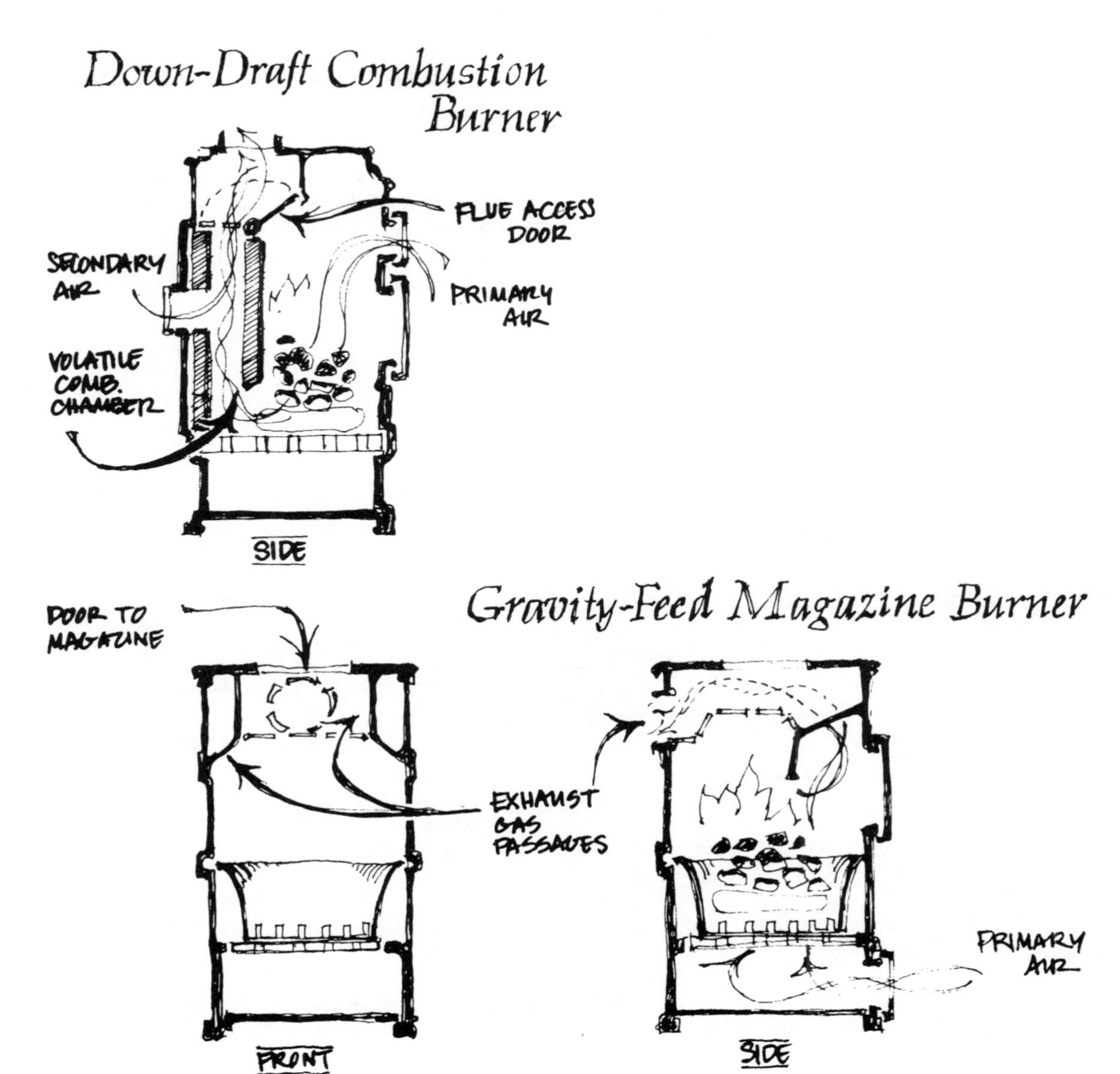

chamber where more of the volatiles are burned. Some models will burn anthracite while others are specifically designed for bituminous coals.

Stoker Systems. There are three types of stoker systems found throughout the United States for central heating and in commercial applications. In each system, small sizes of coal (usually rice or barley) are fed onto a grate where a relatively thin layer of coal is burned at one time. Usually coal is fed to the grate through the use of an auger; in some systems the ash is automatically removed as well. Stokers are designed for specific sizes of anthracite or bituminous coals with very specific burning characteristics; for this reason it is a wise idea to check on the quality and availability of the coal in your area before considering such a large investment.

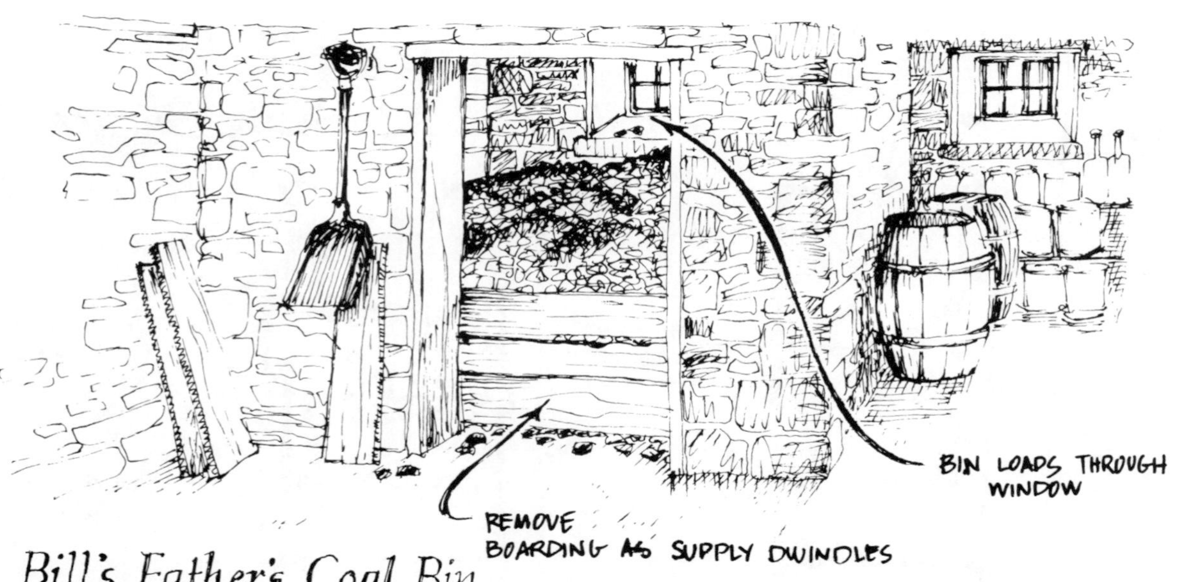

Bill's Father's Coal Bin

Coal Storage and Handling

Back in the heyday of coal burning when most of the homes across the United States were heated by coal, it was a common sight to see horse-drawn coal wagons and later, coal trucks, backed up to basement windows with the coal being shoveled or poured down chutes into large sturdy coal bins. The bins were usually located near the furnaces for easy tending and away from the main traffic flow into and out of the basement. For houses that did not have basements and for houses that coal chutes could not reach, the coal was loaded by hand into sacks and carried to a coal bin located in an entryway or shed attached to the house.

Although times have changed, coal delivery and handling techniques have not. Most of the coal delivery trucks you see throughout New England are modified dump trucks. They generally hold six tons or more of coal that is poured out of a sliding door in the tailgate into a coal chute as the bed is raised.

For a basement delivery where the truck can easily reach a window or door above the coal bin, and where there is adequate clearance above the truck for the body to be raised, this system works nicely, and the process is as simple as having your oil or gas tank filled. Where the truck chute cannot quite reach the window, and an extension has to be used, or the truck body cannot be raised due to building overhangs and tree obstructions, it is not so easy. The alternative is to extend the chute and to assist the flowing coal by using a shovel. A third choice and the most work, is to dump the coal onto a piece of plastic and shovel it into the bin later. (This sounds worse than it is. A ton of coal can be moved easily and quickly using buckets and a wheelbarrow.) If you are fortunate enough to have a coal dealer with modernized equipment, the coal will be transferred into your bin by using a conveyor belt system.

For the less dedicated coal burners and for people who don't have the space for a coal bin, buying bagged coal can be the best option, although it is by far the most expensive.

The best place to store coal is along the inside of your garage or in your basement where it is easily accessible and can be kept out of the snow and rain. Coal stored outside must be protected. If left uncovered, wet coal pieces will freeze together. (In some cases dynamite has had to be used to loosen frozen coal in freight cars.)

Fifty-five gallon drums make inexpensive outdoor coal storage bins. One drum holds approximately 500 pounds, or a ¼-ton of coal. Whatever you choose to use for coal storage, use it only for coal. Never allow combustibles such as old rags, wood scraps, or newspapers to find their way into the bin. Bituminous coals are prone to

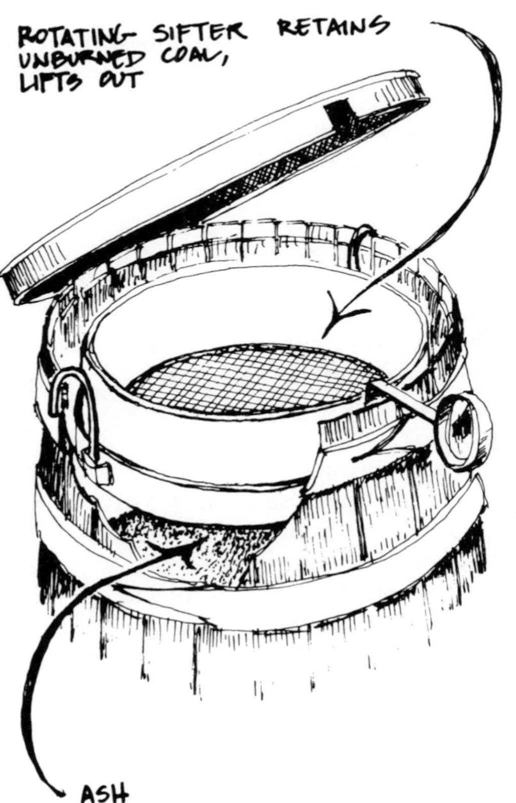

Beacon Coal Sifter
c. 1907
PORTLAND STOVE FOUNDRY CO.

Sifting Clinkers

One of the least favorite chores of the oldest boy in a Vermont household at the turn of the century was sifting clinkers. Back then, the "use-it-up" ethic prevailed over all phases of everyday living, and the fueling of the stoves was no exception. Most families couldn't bear to clean out the unburned chunks of fixed carbon called clinkers that regularly showed up in the ash pan.

Ashes were commonly dumped in the basement until spring, when the entire accumulation from the winter was hauled away. Near each pile was the sifter, a wire mesh screen across a hardwood frame. The coal hod full of ashes was allowed to sit for a few days until the coals were mostly out, and then could be sifted to separate the larger chunks. The clinkers would then be mixed back in with the next fuel load to prevent any potential heat from being wasted.

spontaneous combustion. Large air voids or pockets that form between coal pieces when varying sizes of coal are mixed speed up the slow oxidation process that occurs at the surface of this type of coal. If heat generated by this process is trapped by moisture, dirt, or scrap materials, it can build up to a point at which the coal can ignite. Most of the problems with spontaneous combustion occur in large mine storage piles or in huge bins of bituminous coal where the coal is not handled carefully, but it is something to watch out for. To avoid any chance of this happening, rake the coal down when it is placed in your bin, and be sure to keep the area cool and dry.

Dust is usually not a problem with most anthracite coals, but it can be a problem with the higher volatile anthracite coals (10% volatile or over) and with bituminous coals. To minimize coal dust, design your storage area so that instead of walking into the bin, you reach in with a shovel or scoop. Use a sturdy coal scuttle or hod to carry the coal from the bin to the stove to reduce the chance of spilling the coal.

For some reason cats love to use coal bins as litter boxes. To keep your coal odor-free, keep your cats out! And finally, to keep your woodwork clean, keep your kids out too.

Coal Ash Disposal

Coal ashes aren't good for much, and can produce undesirable effects in the soil if used improperly. In areas where coal-fired boilers are common there is a market for the cindery ash as a base for the paving of roads; in some areas the gravel available may be poorer than the coal ash for such purposes. You won't have enough ash to market, but should you have an existing gravel driveway or path you may use the coal ashes to replenish it, or use them as a base for new construction. They also provide excellent traction on icy walks (be sure you have a good efficient doormat so your floors don't suffer), and a container of *old and thoroughly dead* coal ashes in the trunk of your car may save you the cost of a winter service call. By sprinkling ashes generously in front and in back of spinning wheels you can often extricate yourself in minutes from seemingly impossible situations. Keep a container full for next winter.

Don't use coal ashes in your garden. They can contain dangerous amounts of toxic heavy metals: arsenic, bismuth, chromium, lead, nickel, strontium, and zinc to name a few. Though the mineral content will differ from mine to mine, the risk is not worth taking. Plants will take up the toxic elements as well as the beneficial, and your body will concentrate them in unacceptable places such as the brain, nerves, reproductive organs, and kidneys. Bituminous coal, in addition to the traces of heavy metals, can contain up to 10% sulfur trioxide, highly toxic to plants in those concentrations. A high iron-sulfur coal ash is not only conducive to increased clinker formation, but terrible for growing plants. With this in mind, avoid the use of coal ash in any situation where the rain can leach these undesirable elements into areas used for growing human or animal food. Should you use the cinders for a pathway base, as mentioned previously, be sure the run-off bypasses such areas.

The Stovetop Chef

The "summer kitchen" is a part of the house that you will have trouble finding in homes that were built within the last 50 years. Prior to 1930, though, young women would look with special favor on a suitor who talked of setting up housekeeping in a home that included a summer kitchen. In those days, food often was cooked on a kitchen range that burned either wood or coal, and on sultry summer days the heat in the kitchen could be unbearable; hence, a special porch-like room that was cool and airy just for hot weather cooking.

Heating with wood or coal seems like a primitive exercise to many people, and cooking with these fuels is frequently considered even more unique. It doesn't take a woodburner long, though, to discover that the stove top is an ideal place to incorporate the use of solid fuel to a greater extent into one's lifestyle. Stovetop cooking is energy efficient, convenient, and claimed by many to provide better-tasting food.

Many stoveowners begin their venture into stovetop cooking with a decorative tea kettle. Often purchased for aesthetic appeal, the tea kettle soon lets the cook-to-be know that it has a system of signals that announces the various levels of the fire. A hissing sound means a hot fire that is good for rapid boils or quick frying. A quiet, gentle steam winding from the spout indicates a perfect temperature for stewing and simmering, and no signal at all means that the stove isn't hot enough to cook on yet.

The simmering tea kettle on the stove contributes humidity to the air in the house as well as a constant supply of hot water for tea, chocolate, or bouillon. Our ancestors would keep a special pan of water on the stove to which they would add pickling spices or a potpourri of cinnamon, cloves, and allspice. During the holiday season, sprigs of balsam or pine needles would be added for a more festive aroma.

A second stovetop accessory that is usually acquired is the trivet for the tea kettle. Trivets are decorative cast pieces that hold pans up off the hot surface, usually a $\frac{1}{4}$ to $\frac{1}{2}$ inch. Since many tea kettles do not hold enough water to last the entire night on an airtight stove, they can boil dry before morning and possibly cause damage to the kettle. A trivet placed underneath modifies the burn rate sufficiently so that the water evaporates more slowly.

The trivet is useful for cooking as well to give the cook flexibility in stovetop temperatures. In contrast to the heat level of a modern gas or elec-

Maine's Antique Stove Expert

Bea Bryant has a reputation that has attracted customers and antique stove buffs from as far away as Japan.

For nearly a decade she has rescued parlor and wood cookstoves from back sheds, barns, and junk piles, and offered them for sale in reconditioned shape.

Along with her husband and sons, she owns and operates Bryant Steel Works from the converted family farmstead in Thorndike, Maine. Their collection of vintage stoves is staggering and includes a separately housed museum of 75–100 fossils from the stove industry's evolution.

The company, which has employed as many as 15–20 fulltime helpers, specializes in the restoration of antique cook stoves, both wood- and coal-burning.

The Bryants will spend hours carefully matching a prospective buyer with a kitchen range to fit his or her needs, even if it means entirely rebuilding one from abandoned stoves in their collection. Parts are salvaged and sold from over an acre of graveyard inventory. Each new acquisition is completely disassembled, sand-blasted, repaired, and recemented before being shipped to its new owner.

Interested stove buyers can shop over the phone by calling Bea directly or by making the pilgrimage to the antique stove mecca.

Bryant Steel Works is open for business Monday through Saturday or by special appointment. Visitors traveling to Maine from the south should follow Interstate 95N to the Fairfield exit. At the exit take RT 139 and head east toward Thorndike (approximately 20 miles). After passing through Thorndike village, continue for two more miles, and take the first road to the right, beyond which the way is clearly marked with signs.

tric range that can be adjusted by the turn of a switch, the wood- and coalstove chef must make adjustments by raising or lowering the cooking utensil, or by moving it to a cooler part of the stove. A dish may be moved from a hot stove surface to a trivet when it begins to boil, then returned when it begins to cool. This technique is especially helpful when you are waiting for the reaction to an adjustment of the air control.

Trivets may be used to protect fragile foods, or to tenderize tough ones by making possible a long period of simmering time. A cast iron trivet may be covered with aluminum foil and used to warm up rolls or desserts, or thaw frozen foods without damage.

Most experienced cooks prefer heavy cookware, cast iron or heavy aluminum, which distributes heat better than lighter weight cookware. Food seldom sticks or burns in them.

Cast iron has long been appreciated for its cooking qualities and durability. The thickness of the material absorbs heat, spreads it evenly, and retains it over a long period of time. You can buy new cast iron from manufacturers who never stopped producing it or collect old cast iron cookware with its hand-wrought patina.

Although pitted or cracked pieces should be avoided, old iron can be restored with a little effort and patience. The method is simple. After scouring off the surface rust with steel wool or sand, wash the pot with soap, and dry it thoroughly. If the piece has been painted, remove the paint with a commercial solvent and steel wool. Rub the utensil with unsalted shortening or lard and warm it in a slow oven (200°) for about two hours. Wipe clean with paper towels and continue to repeat this treatment until the piece acquires a smooth surface.

When you purchase a new piece of ironware, it should be washed with hot soapy water, rinsed well, dried thoroughly, and rubbed with unsalted shortening or lard to prevent rust. Grease only the inside of ironware and inside lids. After oiling, a new pan should be placed in a warm oven (200°) for three to four hours, to season. When the iron has cooled, wipe it with towels.

During the seasoning process, the minute pores of the cast iron pieces open, allowing the hot fat in. The fat is burned to form a glaze barrier between the food and iron. Once seasoned, cast iron parts should not be washed with soap. They can be cleaned easily by allowing water to sit in them for a period of time, until the softened debris can be rubbed off with a loosely woven plastic pad. To prevent rusting, about the only thing that can harm an iron utensil, wipe the inside of the dry piece with a lightly oiled towel before storing. With a little extra care, your cast iron pieces will last a lifetime.

Recently, it has become popular to buy enameled cast iron, which has all the advantages of cast iron but does not require seasoning and is easier to clean. This does not necessarily mean that you need to buy new kitchen specialty items or trek to auctions and flea markets in search of the worn and proven. Most of the basic items of your kitchen inventory can be adapted to wood stove cookery.

Long-handled spoons, spatulas, and forks will help you remain at a comfortable distance while cooking on a radiant wood heater. You'll also want to keep a supply of well-insulated pot

The Flagship of Stove Design

The wood-burning range—dependable for warmth, cooking, and good cheer during the cold season—is a complete kitchen unto itself.

It was developed in the early 1800's when "pot belly" parlor heaters were popularly in use. In contrast to the round base burners, the functional kitchen stove provides not only heat, but ample cooking space matched with a beautiful character and elegance.

From the beginning, these cast iron ranges had a large oven for baking, usually under the right side of the cooking surface. Heat comes from the firebox, a compartment where either coal or fuelwood is consumed, and goes around the top or around the oven, depending on the stove model.

Warming ovens, closed boxes above the cooking surface retain some of the rising heat. They provide an ideal place for bread to rise, herbs to dry, or plates to warm before serving.

The stove's surface area can have from one to six removable cooking plates, or fondels, similar to burners on an electric range. Cooking on a stove top can be as uncomplicated as browning toast or as masterful as a stir fry skillet dinner. The temperature of the cook top varies in degrees of intensity, from searing hot to simmering. The hottest spot is over the firebox, where a tea kettle can be left to steam away the "cooking water."

Some woodstoves also have a reservoir designed to keep rain water warm; or, in many cases, a copper-lined reservoir that preheats the domestic water supply.

Besides performing its utilitarian functions, a big wood stove in the kitchen is unrivaled for dressing in front of on cold winter mornings, drying soggy hats and mittens, and making the best of green stovewood.

Blackened and nickel-bright, it evokes old-fashioned memories and draws us to its gleaming presence.

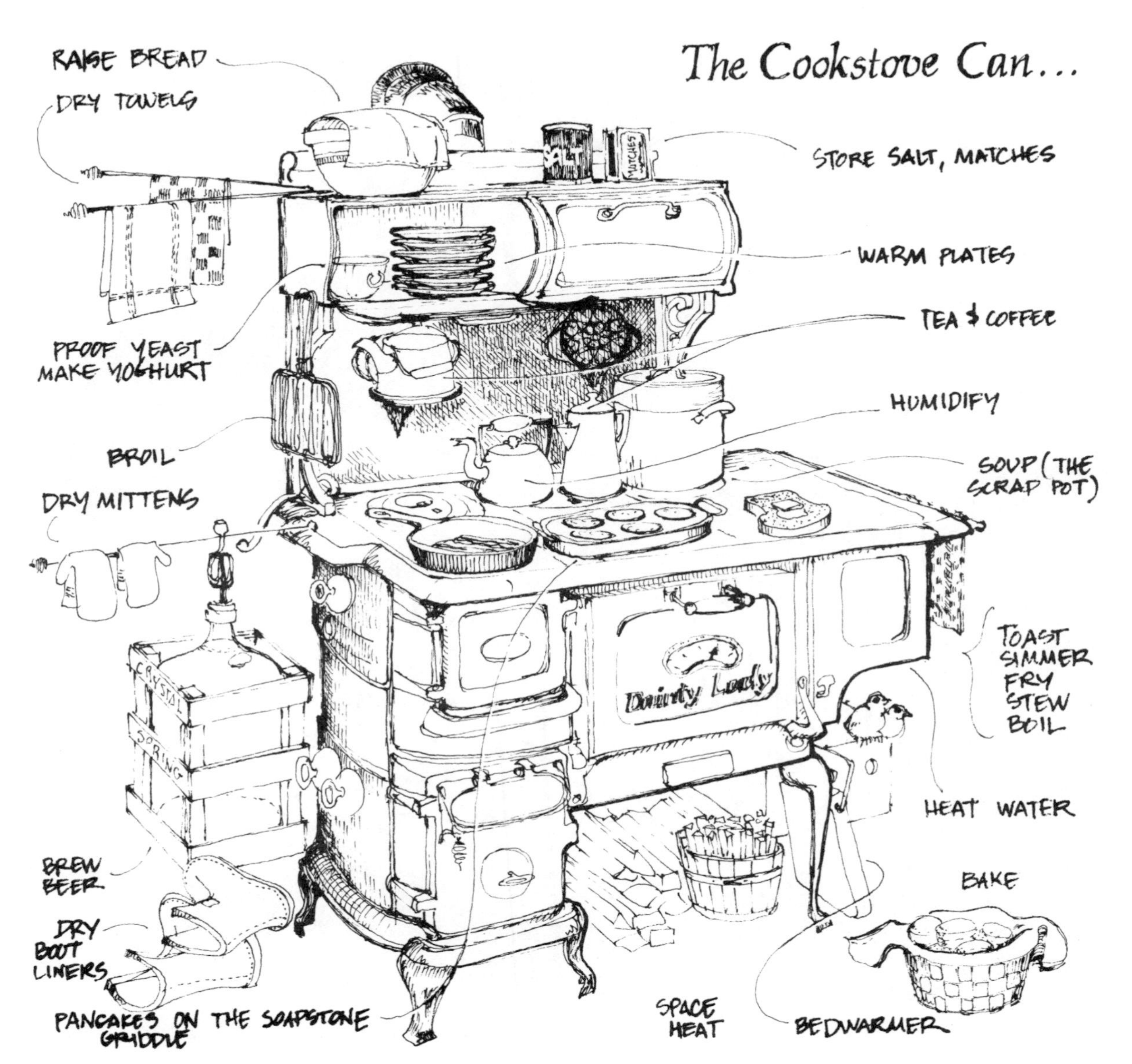

holders and mitts nearby to adjust the draft and handle pots.

Although our grandmothers used less conventional methods to test the griddle surface temperature, a wood stove thermometer, made with a magnet to hold it in place, is designed to eliminate the guesswork. It not only tells you at a glance the temperature of the surface, but, by comparative readings, whether the stove is maintaining even heat or getting hotter or colder. The stovetop temperature is not as critical for foods that can be cooked for a long period of time over low heat as for those that must be cooked for shorter periods at higher temperatures. However, by recording appropriate cooking temperatures, you will find it easier to duplicate successful recipes. The cooking guide given below will help you to translate the stovetop thermometer's readings into heat requirements for cooking:

Very Low	250–300°	(slow cooking)
Low	300–350°	(simmering)
Moderate	350–375°	(stewing, baking)
Moderately Hot	400–425°	(frying)
Hot	425–450°	(boiling, browning)
Very Hot	475–500°+	(quick fry)

One of the greatest challenges of wood stove cookery is learning to allow the proper length of time necessary to cook certain foods. Several factors, including your choice of fuel, will affect cooking times. Any type of seasoned wood, which is drier and lighter, will provide a more dependable and hotter fire. Under ideal conditions, the type of wood you choose will be matched to the type of cooking you wish to do. Hardwoods of maple and oak provide steady, hot fires for baking. Elm, birch, and ash give long moderate fires for slow-cooking dishes. Pine, poplar, and birch burn quickly and hot for a responsive fire. Hickory and fruitwoods not only burn well, but release a pleasing aroma as well.

There are no specific guidelines for reaching and maintaining a particular heat level. Airtight, cast iron stoves, by their very nature, will hold a steady heat over a long period of time. With practice and observation, you will learn to regulate the intensity of heat output by adjusting the air control settings, and the internal damper, if your stove has one. These two controlling devices control the rate of the air entering and leaving the combustion chamber. The more oxygen presented to the wood, the faster and hotter it will burn.

An established fire with the damper open will generate a high heat, between 400–500°. By closing the damper, the temperature will drop to between 250–300°, depending on where you set the thermostat. If the temperature drops below 200°, open the damper to rekindle the fire. If the stove gets overheated (above 500°), close the damper and thermostat to help cool the fire.

Try cooking directly on your stove's griddle if it has one. Slices of homemade bread or English muffins can be toasted to a golden brown in less than five minutes. A freshly kindled morning fire will provide a griddle temperature just right for frying eggs or browning pancakes. Also, experiment with grilled sandwich combinations.

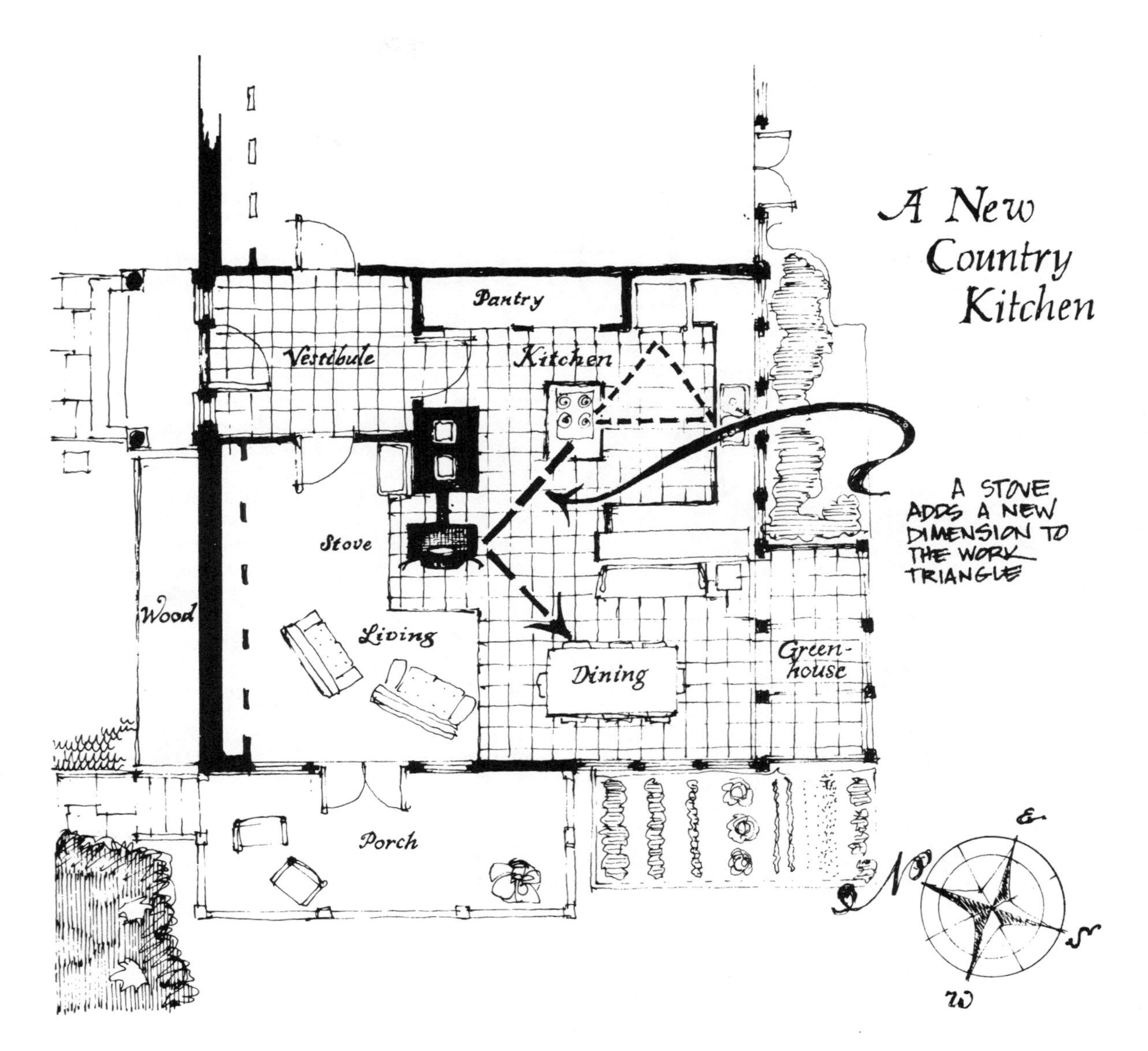

The Stove Cosmetologist

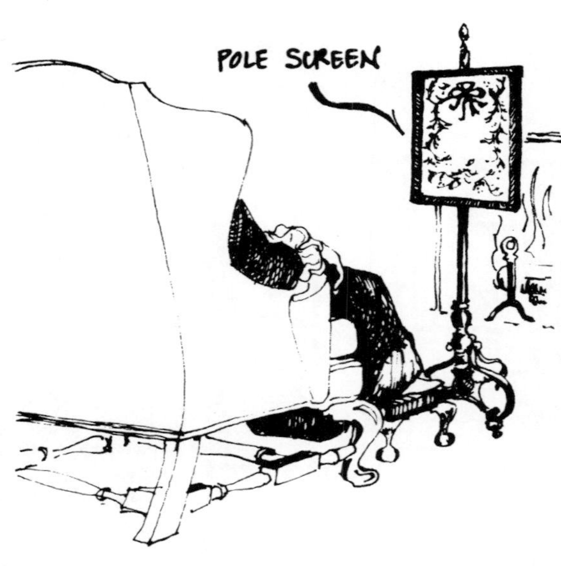

There is something charming about the image of apples and herbs hanging about the colonial fireplace to dry. The sense of enchantment fades when we realize that moisture is leaving our skin as well when a wood or coal stove is heating our house.

Wealthy women in the 18th century were very familiar with the effects of fire heat on their skins. It was fashionable to wear makeup at that time, and some of the preferred lotions contained a lead base. Cold winter evenings required a seat close to the open fire, and the radiant heat from the flames could react with the lead-based makeup to cause uncomfortable irritation to the skin. Since neither sitting a greater distance from the fire nor parting with contemporary fashion were acceptable alternatives, the women employed a special upright frame of embroidery placed near the chair to shield their faces from the fire.

In modern homes, wood and coal heat creates a dry atmosphere that can conspire with wind and chill of winter to cause dry, chapped skin. You can combat winter dryness in several ways. First, get more moisture into the air. An electric humidifier is the most effective, but a container of water on the stove surface or under the stove will help too. Dry your washed clothes indoors on a clothes rack, and you will save money on your utility bill at the same time.

A popular story about an unusual Hollywood beauty treatment describes putting one's face under a towel-tent over a pan of steaming water into which rose petals are dropped. Most stove tenders will resist the prospect of bending over their hot stoves and few have rose petals on hand in the dead of winter, but there are some specific treatments that will bring relief. A light application of vegetable oil applied to the face after a thorough cleaning will preserve and offer good skin protection. Wheat germ oil, with its high vitamin E content, is especially good. Mineral oil, however, is very drying and should be avoided.

Hair will also benefit from a conditioning treatment of vegetable oil. Rub two tablespoons of oil into wet hair concentrating on the scalp. Wrap the hair in a hot towel for 20 minutes, then shampoo thoroughly.

Children especially will enjoy popping corn over the hot surface in an old-fashioned, long-handled popper.

A stove griddle is made of the same material as a good frying pan and should be treated as a new cast iron piece. Curing (seasoning) should be completed before the stove is first used, and cleaned after each use with paper toweling. Once the griddle has been cured, do not wash it with soap and water. With use, it will darken naturally to match the rest of the stove.

Fall Checklist

By now it is too late to accomplish much. If your heavy installation work is not already completed, you are in big trouble. Your neighbors who spent Mud Season making fun of you are now lined up waiting for coal deliveries and the chimney sweep. A few gaze forlornly at your woodpile and ask if you have any extra to sell. Don't rub it in by inviting them over to watch the World Series when you know they have to cut wood.

- □ ***Sell that clunker.*** *Assuming you were forward-thinking and solvent enough to buy your new unit while the off-season specials were in effect, you now can unload your old unit during the height of the season when buying fever is at its highest pitch. Break out the wire brush and touch-up paint, and transform that rust bucket into a cream puff.*
- □ ***Take a foliage trip to Vermont.*** *Enjoy the season. You worked hard in the spring to earn the right to relax now.*
- □ ***Final logistics check.*** *The 50-foot walk from woodpile to stove which does not seem far at all in August will be like a trek to the top of Everest during a January blizzard. Load cellars, porches, and woodboxes to the gills, then as you start burning your stove, do not use this easily accessible wood. That's for when it gets nasty out, remember?*
- □ ***Buy briquets.*** *Charcoal briquets are terrific for starting coal fires, but when you try to buy them in January, people at the store will think you have a screw loose. Buy now. Top off your kindling supply, too. Do not use self-starting briquets or flammable starting fluid. Both can be dangerous.*
- □ ***De-birdproof your installation.*** *This is an easy thing to forget until the smoke pouring back into your house reminds you of your creosote-clogged efforts of several months earlier.*
- □ ***Buy a punching bag.*** *At some point just prior to lighting your stove you will find yourself in need of 13 inches of 6-inch pipe. You have only a 12-inch piece, and the hardware store a 24-inch piece. You will then proceed to cut the pipe to 12¾ inches and your finger as well. Rather than throw the pipe through the window, thereby compounding your homeowner problems, punch the bag.*
- □ ***Doublecheck stovepipe connections.*** *Summer moisture can rust elbows. If you worked on your installation, did you reattach stovepipe sections with sheet metal screws?*
- □ ***Recharge your fire extinguisher.*** *That unit has been sitting unused for many months. While you are at it, test your smoke alarms, and replace batteries if necessary.*
- □ ***Start it up gently.*** *With a season of burning under your belt, the tendency is towards overconfidence. Even veterans sometimes break their stoves on the first fire of the new season because they have ignored the rule to never build a hot fire in a cold stove. Follow a break-in procedure similar to that used when you first bought the stove.*
- □ ***Oops!*** *Did you remember to remove the cooking oil from the griddle? If not there is now an acrid smell in your house. Use a mild detergent and water, but most importantly, dry the griddle thoroughly as rust will set in quickly. (P.S. Don't forget to remove the kitty litter.)*
- □ ***Open the windows again.*** *Review your springtime operating techniques. they will work just as well now.*
- □ ***Make sure the woodpile is securely covered.*** *An early 6-inch snowfall can make your life miserable. Old roofing tin or canvas tarps make good covers, but plastic will do. Use old tires to hold cover in place.*
- □ ***Take the spouse and kids to stove school.*** *You might be away at work for most of the day, but someone will have to tend the stove and to make sure the kids do not hurt themselves. The entire family is heated by the stove, so the entire family should know how to operate the stove. give the babysitter a lesson, too.*

4

Winter

The Big Lie
Advanced Stove School
Doctor of Stoves
The Muse of Stoveland
Winter Checklist

Stripped of its shroud of foliage the land seems naked and honest. Survival is a chore, but mastery of the elements with the aid of skis and stoves makes one feel superhuman. The real test is not one of extremes, but rather of endurance. Can you believe that come April you will be glad to see mud?

The Big Lie

The calls come in like clockwork. For months, since the previous March, our customer contact has been largely beneficent. We have been working with people as partners in an exciting venture to help them save money through a home improvement which is as beautiful as it is practical. We assist people with questions on sizing and stove placement. We advise them on air movement, installation, hearth construction, coal purchasing, and wood stacking techniques. We're comrades, united in the quest to live in a world of simple and honest values.

Cracks in the veneer become apparent in mid-September. People who should have been planning for the winter months long ago are now starting to panic. They call needing a stove yesterday. They want their wood stacked in the cellar, but they have not yet bought the chain saw. They put an edge on the most beautiful season because they remind us that cold weather is stalking us. And time is growing short.

The first crisp weekend in October the calls are predictable, such as what are the proper break-in procedures, requests for small parts from those who have been postponing the reality of life with their stoves, and minor trouble-shooting. By now the leaves are fallen, and the earth is dead in expectation of Winter. It might as well snow, as the ground seemingly begs to be whitewashed.

After the first cold night of the season the deluge begins: "My stove won't heat,"—"my stove won't burn overnight,"—"my stove smokes" . . . we expect and receive them all. A lady in Connecticut ran her stovepipe from stove to upstairs bedroom, and no farther, expecting the smoke to miraculously disappear, perhaps through an open window. A man on Long Island never removed his ashes, expecting that "someone from the stove company" would stop by to do it. The problems are as unique as the individual callers and as timeless as the art of stove burning.

If there is a message to be learned from this book, it is that owning and operating a stove is a year-round venture. Somehow it seems fitting that the good Samaritan who prepared his fuel back in Mud Season, tended to his installation in Summer, and worked out the bugs in Fall would have earned a free pass for Winter. This person should be able to dial the stove to "cruise control" and sit back in the easy chair reading *War and Peace* while the blizzard rages outside. Alas, life

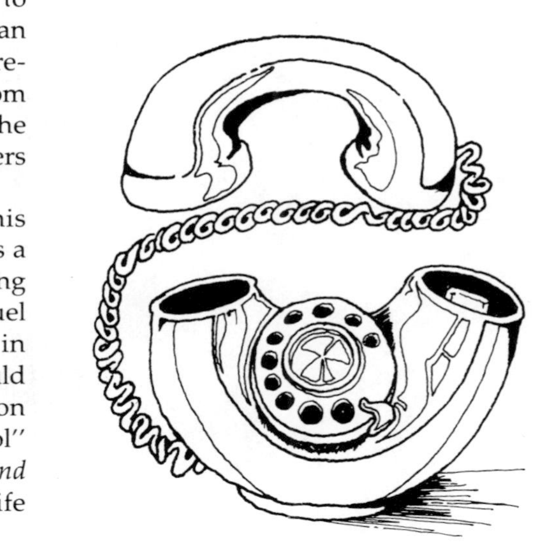

is never just. In fact, the season of hauling wood, coal, and ashes has just begun. For the next few months you will be an indentured servant to your stove.

The self-loading, self-cleaning stove has not yet been invented. Stove ownership, like having a child, is a choice that involves commitment. Simply stated, you should not have one unless you are prepared to change a few diapers. And as with a child, there will be many times when you will question whether or not the end result is worth the effort.

Why then, bother? Certainly there are financial benefits, but if you dedicated your hours spent cutting, splitting, stacking, and hauling to, for instance, pumping gas parttime at the local station, you would certainly be able to pay your oil, gas, or electric bill and still have money to play the video games at your local bar. No, financial benefits are a pleasant bonus, but not the real motivating factor in stove ownership.

There is an aesthetic reason to own a stove, but the beauty of a stove is meaningless unless it is married to function. This leaves an even less tangible reason: You like it. A stove brings you into contact with your environment, just as athletics bring you into touch with your body. The weather takes on new meaning, as do the trees.

The Customer Relations staff members at Vermont Castings have become adept at coping with the folks whose romantic illusions of stove ownership suddenly have confronted the reality. There is nothing quaint about a puddle of creosote or a scorch mark in a rug from a careless coal. But if we can guide our customer past the first panicked call, we have a friend for life.

Advanced Stove School

This chapter is your senior year. Stove School is old-fashioned in that the goal is to graduate students whose store of knowledge is well-rounded. You already know about fuels, you own a stove or know what you want in one, and you have successfully passed the installation test.

This section contains science and philosophy, at least to the extent that these formidable disciplines pertain to burning wood and coal. You already know how to build a fire; the next few pages will tell you *why* the fuel burns. We will look at why wood ashes are good for the garden and why coal ashes are unacceptable. We will also tackle the ethical question of pollution, and to what extent the wood or coal you burn in your stove contributes to this international concern. And finally, we will look at the future to see how stoves and fuel will be developed to satisfy the dual demands of higher efficiency and minimal environmental impact.

Combustion Technology 101

Combustion, in general, is defined as the rapid oxidation of elements or compounds that liberate heat during their combination with oxygen. Fuels (solid, liquid, or gaseous) that are commonly used are those containing a high proportion of elements or compounds which readily oxidize. Although we are mainly concerned here with solid fuels, the basic combustion theory is independent of the fuel classification. All common fuels are composed of the basic fuels, hydrogen and carbon, plus varying amounts of other elements.

Wood Combustion

In wood, the basic fuels are locked up in complex molecular structures (which also include other elements) within the lignin and cellulose of the wood. (Cellulose can be thought of as a wood fiber and lignin as the glue that bonds the fiber.) Wood can also contain a high percentage of inherent moisture that, although obviously not combustible, can be a significant participant in the combustion process. Wood combustion can be separated into the following processes:

1. *Evaporation of Inherent Moisture.* Moisture is evaporated from the surface of the wood as the temperature of the wood climbs above 200°F. Energy consumed during this drying process must be supplied by burning wood or kindling.

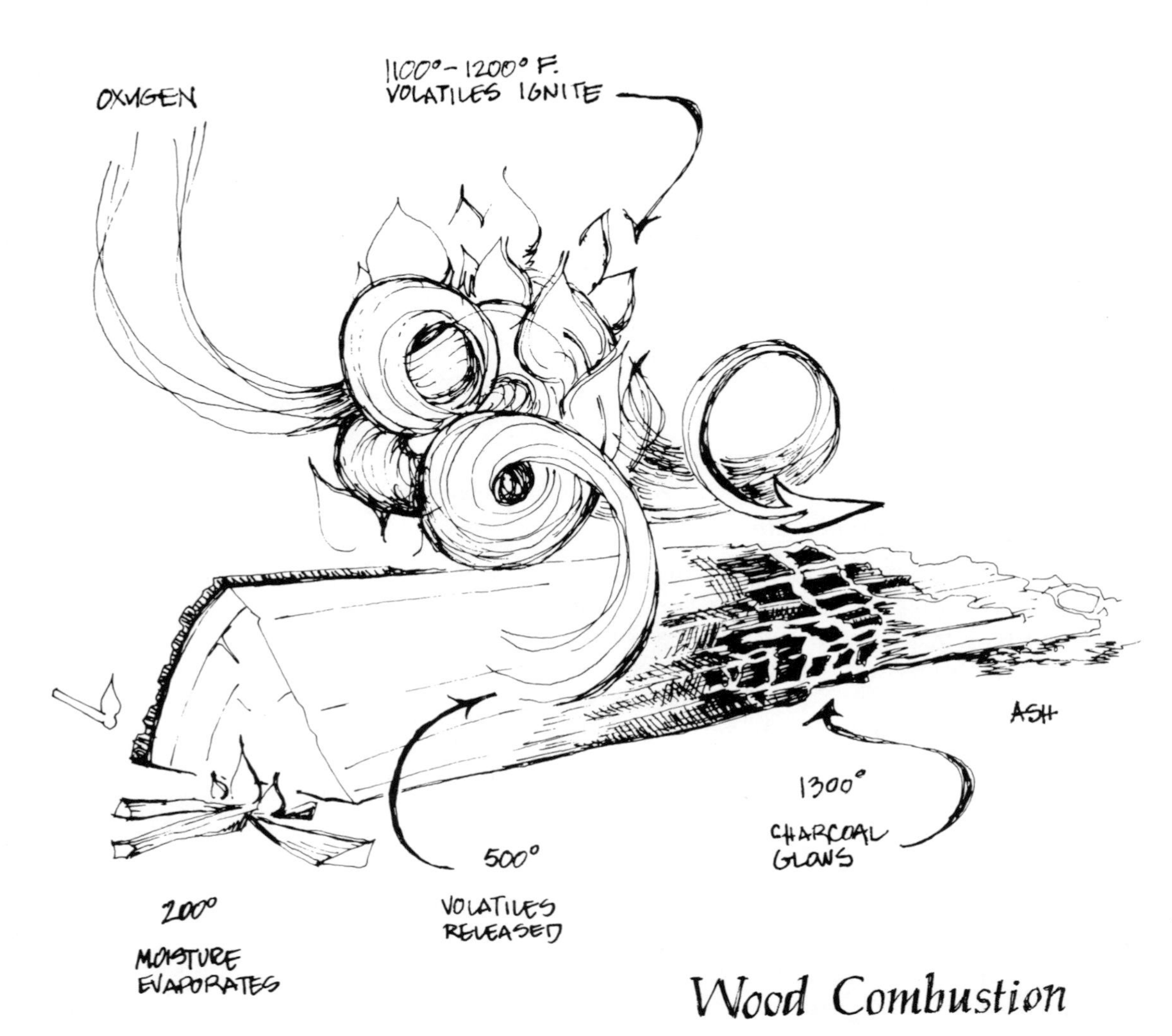

The moisture in the wood controls the rate of this process. The higher the moisture content, the more energy required to dry it, and the slower the rate of initial combustion.

2. *Release and Combustion of Volatile Materials.* A large percentage of the wood's energy, one-third or more, is contained in the gaseous and liquid/vapor materials released as the temperature exceeds 500°F. These gases and liquids (tar vapor and droplets) are the products of distillation (pyrolysis) of lignin and cellulose and contain molecules composed of various combinations of carbon, hydrogen, and oxygen. Release of these energy-laden materials does not insure that this energy will be useful. Only combustion will release the energy. This requires that two conditions be satisfied: a) sufficient oxygen must be present and well mixed with the volatile gases, and b) the temperature in this air/gas mixture must be sufficiently high to cause ignition (i.e., 1100–1200°F). When both conditions are satisfied, the result is the flaming combustion observed above the surface of the logs.

3. *Charcoal Combustion.* When the volatiles have been driven from the wood, the material that remains is charcoal which is made up mainly of carbon. Charcoal can contain as much as two-thirds of the wood's energy. The combustion of charcoal occurs at its surface and is recognized by an orange glow. The glow indicates the combustion of carbon mixed with oxygen at elevated temperatures (1300°F plus) to form carbon dioxide. Ash remaining after the charcoal is fully consumed is mainly inorganic material (minerals, etc.) that cannot be vaporized or combusted.

The Reality Factor

Unfortunately, in the world of the woodstove we all know and love, the process of combustion is more complex than the three distinct processes just mentioned. Although all these processes are occurring, the complicating factor is that all occur concurrently rather than consecutively. Also, except perhaps at the very beginning or the very end of a fuel load, we can never be sure which portion of the fuel is involved in which process. Quite likely in a given fuel load, one log may be drying while another is pyrolyzing while still another is burning as charcoal. Even more likely is that any single log is, to some degree, involved in all stages at the same time.

This variability makes wood-log fuel combustion one of the most difficult combustion processes to understand and to analyze. It may be noted that some of the variability associated with wood as a fuel is reduced when smaller pieces such as wood chips or pellets are used. With an understanding that many things are occurring simultaneously within a woodstove, one can look in more detail at the factors that affect woodstove performance:

Inherent Moisture

The old adage that too much or too little of anything is not good applies especially to the moisture content of wood. Moisture in wood, although not combustible, can dramatically affect the combustion process. The presence of too much water (typically 30% to 40% moisture content for green wood) can consume so much heat as it is boiling away that not enough heat is left to sustain or initiate ignition, or to create the elevated temperatures necessary for good combustion.

If the wood is too dry (less than the 10% moisture content which develops if wood is stored indoors for long periods of time), equally bad things can happen. The wood can pyrolyze so fast after being kindled that the volatiles released can exceed the capabilities of the oxygen present to combust them.

Thus, wood that is too wet or too dry can cause similar results. Translated into practical terms for the woodburner, temperatures too low for combustion or inadequate oxygen both can result in loss of efficiency, high organic emissions up the stack, and a possible creosote hazard. Fortunately, properly seasoned wood with a moisture content of 15% to 25% minimizes either potential problem.

Volatile Combustion

The volatile material released during wood pyrolysis is quite difficult to combust. As it is released through the surface of the wood, it must be well mixed with the proper amount of oxygen; this mixture then must be heated to the ignition point of 1100–1200°F. If either condition is not met, combustion of these gases and liquids/vapors will not occur, and the result will be smoke (the aerosol form of tar droplets), potentially hazardous creosote formation, *and* the loss of a substantial amount of the wood's heating value.

Several factors can make these volatiles difficult to combust in typical wood-burning appliances. When air is limited to control combustion rate, the available oxygen above the fuel bed for combustion of volatiles is reduced. This lowers velocities and, consequently, the mixing in that zone. Also, the temperature must remain high until volatiles are fully combusted, a difficult feat in conventional wood heaters where a basic design criterion is good heat transfer away from the combustion zone. The combustion of these gases can be accomplished, though, at medium-to-high firing rates. This emphasizes the importance of properly matching your stove to your heating needs.

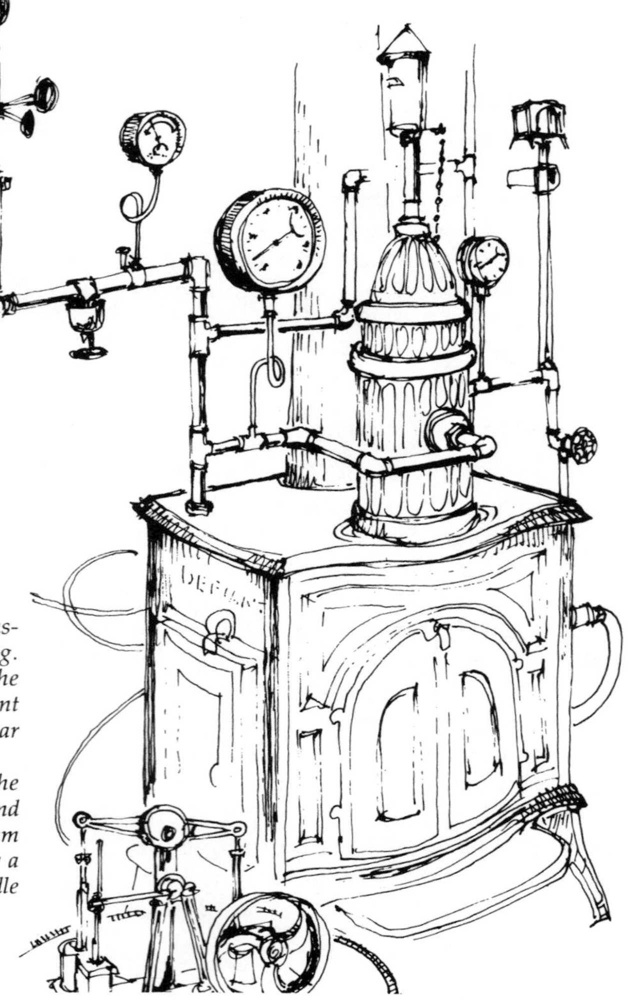

The Steam Defiant

A steam-powered Defiant greeted the 10,000 customers who showed up for the 1981 Owners' Outing. The creation was the product of a blitzkrieg by the Research and Development department of Vermont Castings to prepare something suitably spectacular and frivolous to greet the crowds.

Finishing touches were being applied even as the first visitors arrived. After substantial cranking and stoking, the flywheels began to turn and the steam whistle spread its hoarse call across the field. Why a steampowered stove? Well, why not? Of such idle exercises are great enterprises born.

Charcoal Combustion

If adequate oxygen can get to the surface of the hot charcoal, combustion is highly probable. The actual combustion process involves a number of intermediate steps where carbon dioxide, carbon monoxide, and other more complex carbon and oxygen-containing compounds are formed and decomposed before finally leaving the combustion zone.

If a specific stove has tendencies to form deep beds of charcoal without adequate provision for supplying air to the entire charcoal bed, the lower portions of the bed can smother. Although bothersome, this charcoal does not have to represent lost heating potential as it can be separated and recovered when cleaning out the ashes and burned later.

Combustion Air Distribution

From a combustion point of view, a properly-designed conventional woodstove is one that includes features that optimize the utilization of the fuel's energy. Because a substantial portion of the wood's energy is contained in the volatile materials that evolve during wood pyrolysis, and because these materials are much more difficult to combust than the remaining charcoal, it is reasonable to concentrate stove design on the combustion of volatiles and to accept a compromise for the burning of charcoal. This would result in a stove with provisions for both primary and secondary air. Of the total air flow to the stove, a good rule of thumb is a distribution of 25% (20–30% range) primary air and 75% (70–80% range) secondary air. This should transfer volatile materials from the primary combustion zone with adequate secondary air to combust them.

You can recognize volatile combustion by the flaming above the fuel mass, or in a secondary combustion chamber if your stove has one. (Constantly opening a door to try to catch a glimpse of flaming is not recommended, though, since this will lower the temperature of the combustion chamber.) Another good indicator of stove operation is the density of the smoke exiting the chimney. Little or no smoke coming out of the stack means that the volatiles are not escaping unburned.

Anthracite Coal Combustion

Anthracite coal is compact and has a high heating value. Composed primarily of carbon (80% or more), its combustion is similar to charcoal in that it is primarily a carbon-oxygen reaction at elevated temperatures.

Depending on the type of stove, actual zones where distinct combustion reactions occur can be defined. Coal combustion is much more clearly understood than wood for several reasons: Coal is nearly a single component fuel (carbon), and once a coal fire is established, a consistent chemical reaction continues throughout most of the burn

The Wood-fired Zamboni

In Warren, Vermont, there is a small pond which makes a perfect hockey rink. A perfect hockey rink, however, needs more than a great pond. Cold temperatures are required, as well as a means of ensuring a smooth surface. The ideal piece of equipment for this purpose, called a Zamboni, can be seen between periods at professional games. Unfortunately, such a piece of gear is well beyond the means of your average group of counterculture back-to-the-landers in Central Vermont.

Yankee ingenuity prevailed. Two fifty-five gallon drums were combined with a discarded hot water heater. Someone scrounged some steel tubing, and after some furious cutting, banging, and welding, the Incredible, Amazing, Awesome, Wood-Powered Zamboni was born.

The I., A., A., W-P., Z. (as it was known for short) served its masters well for several years. It came complete with steam whistle and hot water dispenser for the hot chocolate. It was durable. (It even survived a fall through the ice one day when the enthusiasm of the skaters outstripped their common sense.) But, alas, time passed, the hockey players dispersed, and the Zamboni fell on hard times.

Some members of the Vermont Castings Research and Development team heard about the Zamboni and now are restoring it to its former glory. They meet periodically, elbow grease and welding torches in tow, and amidst a flurry of sparks pump life into the rusty hulk. By next winter, with a little luck and enough beer, the Randolph skating rink will have the smoothest surface north of Boston Garden and south of the Montreal Forum.

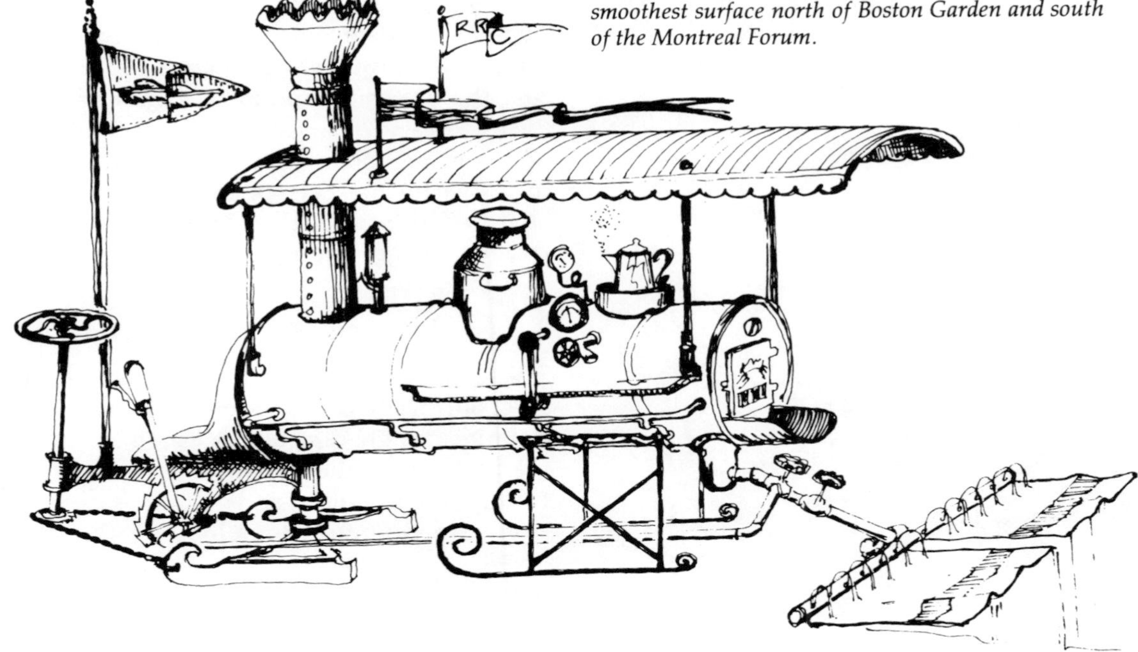

cycle. Also, a considerable amount of coal combustion analysis and experimentation has gone on at the industrial level, much of which translates to smaller applications. And finally, the smaller pieces used in coal heaters contribute to uniformity of the combustion process.

Practically speaking, coal can be a difficult fuel to ignite. This is due to the low volatile content, (10% or less), high carbon content, and density. Whereas wood might ignite at 400–500°F, anthracite will not ignite until 900°F or more. Once a coal fire is established, though, there is little to do but sit back and enjoy the glowing heat and to observe the gentle rolling blue flames above the fuel bed. (These flames are the result of carbon monoxide, which is formed inside the burning fire mass, combusting to carbon dioxide at the surface as additional oxygen is available.)

An interesting difference to note between general woodstove and anthracite coalstove designs is the primary/secondary air split. They are just the opposite; anthracite burns best with a 75% primary/25% secondary distribution. This split can be obtained either by adding separate air above the fire-mass or by keeping the burning coal bed thin enough to ensure that adequate oxygen will get through as secondary air.

Bituminous Coal Combustion

These coals are less desirable for residential use than anthracite coal and their combustion poses definite design difficulties. There are about as many types of bituminous coals as there are bituminous coal mines, and properties vary widely from mine to mine. Bituminous lies somewhere between wood and anthracite as far as carbon/volatile percentages. In general, stoves that burn bituminous coal have about a 50/50 split between primary and secondary air. Special care needs to be taken when using bituminous coal, as pockets of volatile and potentially explosive gases can build up in stagnant areas within the stove. These could ignite abruptly and cause potential hazards. Also, large amounts of soot can build up quite rapidly in stove and chimney passages, causing a blockage or fire hazard.

Biomass Fuel Combustion

A variety of fuels may be available as alternative fuels for the future. They are designated as "biomass fuels" here, indicting a recycling of waste material suitable as fuel. This group includes wood chips, hogged wood (a special, larger-sized wood chip), pellets, and other processed fibrous organic refuse. They usually require special designs or modifications to be burned in a stove and are mainly limited to furnace applictions at this time.

Calculating Stove Effiiciency

Combustion efficiency is a confusing concept. There are different definitions and various test procedures to measure it. The most acceptable definition of "combustion efficiency" is a measurement of the completeness of combustion of the fuel burned. This is determined by measuring and calculating the amount of heating potential escaping up the stack as non-combusted fuel, and comparing it to the amount of heat that would be generated if the fuel were fully combusted. It can be represented by the expression:

% Combustion Efficiency =

$$100 \times \frac{\text{Heating potential of wood placed in the stove as fuel} - \text{Heating potential lost up stack from uncombusted fuel}}{\text{Heating potential of fuel}}$$

"Overall efficiency" is a slightly different concept, and may be considered as the measurement of the total effectiveness of a stove; it can be determined by several methods. The first measures the total amount of heat going up the stack in all forms and compares it to the heating potential of the fuel placed in the stove. This approach, called the *Stack Loss Method,* indirectly determines overall efficiency. The expression

$$100 \times \frac{\text{Heating potential of fuel} - \text{Total heat lost up stack}}{\text{Heating potential of fuel}}$$

represents the Stack Loss Method for determination of the overall efficiency (in percent).

The expression

$$100 \times \frac{\text{Heat into room}}{\text{Heating potential of fuel}}$$

represents the second approach, called the *Calorimeter Room Method* for determination of overall efficiency (in percent).

A third method is the *In-Situ Method* which determines the average heat usage of an actual home on a degree-day basis, using a conventional heating source (gas, oil, electric) and then determines the amount of the alternate fuel required to heat the same home on a degree day basis. The comparison of the two heating requirements gives an overall seasonal average efficiency for the alternate fuel appliance. This method adds realism to the conclusion but requires long periods of measurement. The expression

$$100 \times \frac{\text{Heat requirement of house per degree day*}}{\text{Heat potential of alternate fuel used per degree day}}$$

determines the overall efficiency using the In-Situ Method.

"Heat transfer efficiency" is a measure of the stove's ability to transfer heat to the room from the combusted fuel. It is usually calculated after both overall efficiency and combustion efficiency have been determined from the relationship

$$\text{Overall Efficiency} = \text{Combustion Efficiency} \times \text{Heat Transfer Efficiency}$$

or

$$\text{Heat Transfer Efficiency} = \frac{\text{Overall Efficiency}}{\text{Combustion Efficiency}}$$

The single most important term in understanding a stove's performance potential in the home is *overall efficiency*. This determines how many cords of wood your stove will burn to heat your house (for an average heating season). Typically, a conventionally-designed radiant airtight woodstove has an average overall efficiency of 50% to 60%. This means that the energy from about one out of each two logs you put into your

(* as determined from conventional heating source)

Learn from the "Experts"

The solid fuel home inventor should know what he is doing to avoid having any of his experiments turn into a complete washout.

Duncan's Hot Water Experiment

"*Thinking myself the complete stove wizard, I fabricated a hot water attachment for a 1928 Montgomery Ward coal-burning basement furnace rescued from the junk yard. I had found an old sidearm style heater about 14 inches square made with many rows of copper tubing. I attached it to the top of the furnace and added piping for the water. The idea was to set up a gravity feed system with a water tank installed in an attic loft. Using a garden hose, I filled the 35-gallon tank to within an inch of overflowing. The stage was now set for the maiden voyage. I fired up the furnace, ready for a hot shower. The old furnace leaked like a sieve and ate wood by the armful, but soon the water began heating up.*

Water expands when heated. Soon I started noticing drips from above. The water in the tank had started to overflow, showering our one and only piece of antique furniture. As I moved it to a dryer spot, the copper pipe rattled ominously. There were so many turns and elbows in the tubing, water could not circulate fast enough and a flash boil was taking place. The pressure generated by the pent-up steam blew a fitting, releasing clouds of steam and water. Thirty-five gallons of water suddenly ran into the stove, extinguishing the fire and flushing a four-day accumulation of ashes out on to the oriental carpet. The stove never was used again, and my enthusiasm for heating hot water was at least temporarily dampened."

Duncan Syme

Table 4–1

Final Form	Expressed as % of total ash: Wood	Coal
Silica Oxide $(Si)_2)$	3–15	55–57
Alumina Oxide (Al_2O_3)	0–6	31–38
Ferric Oxide (Fe_2O_3)	50–60	0.5–1.5
Calcium Carbonate $(CaCO_3)$	1–13	—
Magnesium Oxide (MgO)	4–15	0.5
Manganous Oxide (MnO)	1–4	0.5
Phosphorous Pentroxide (P_2O_5)	1–5	0.5
Titanium Oxide (TiO_2)	trace	1.2
Sulfur Trioxide (S_3)	1–3	1–3
Potassium Oxide (K_2O)	5–15	0.5–1.3
Heavy metal oxides	trace	low to moderate levels

stove reaches your house as heat. Even if two units are rated at the same BTU output (heat output capability), the stove with the higher overall efficiency will require less wood during the heating season, a factor which is certainly important in the economics of wood heat.

A Wealth of Ashes

If a random sample of people across the country were asked to list the 100 topics they would like to know more about, few lists would include wood or coal ashes as possibilities.

A person's appreciation of the world depends on his understanding of it, however. The international traveler in a foreign country will enjoy his visit more if he speaks the native language, and the Vermont tourist will appreciate the dairy industry of this state to a greater extent if he knows that brown cows are Jerseys, and black and white ones are Holsteins, and that the two yield different amounts of butterfat.

The same principle applies to ashes. Although relatively uninteresting to look at, a bucket of ash is a cornucopia of ingredients, each with a different chemical composition. An evaluation of the ash from your stove can give you clues to the performance of the stove and reflect your operating techniques.

The reason wood and coal fuels release energy in the form of heat is because during their formation, solar energy was transformed by photosynthesis into a chemical form, basically organic hydrocarbon compounds. In the case of coal, this transformation occurred eons ago when rotting swamp vegetation was compressed under layers of sediment. Wood represents a more recent conversion. The energy stored in these carbon compounds is released through a process of oxidation, either slow (decay) or rapid (combustion). If these fuels were pure carbon and hydrogen there would be no residue at all; the by-products of complete oxidation would be energy in the form of heat, carbon dioxide, and water vapor. But coal and wood are not pure hydrocarbons. They contain both inherent and extraneous impurities. Inherent impurities are those minerals that were part of the vegetation itself (plant or tree), and in the case of coal, sediments carried into the coal-forming swamps. Extraneous impurities come from the mining or logging operations; clays and shales from the roof and floor of the coal seam, and dirt that adheres to the bark of trees. Extraneous impurities from mining often can be removed by washing, though the coal will be more expensive if so treated. When these impurities are subjected to the heat of combustion and adequate oxygen, they too are oxidized. No heat is generated, however, and they are not vaporized. They settle out as ashes. Thus ashes are basically composed of mineral oxides, with a few whimsical touches such as a carbonate here and there. Wood and coal ashes are made up of many of the same oxides, but the proportions are much different. Compare the two by looking at Table 4–1 that lists oxides from complete combustion of a high quality coal and a wood sample.

The ranges of the various compounds are wide. Coal impurities reflect a wide range of conditions present at the time of formation and mining, and wood impurities reflect the elements present in a particular soil and the amount of debris adhering to the bark. The general trends, though, are constant. From each ton of coal used, expect approximately 200 pounds, assuming a 10% ash content and complete combustion. That's roughly seven to nine bushels. From each cord of wood, expect 50–60 pounds of ash.

Coal Ash Secrets

Look closely at your ashes. They are visual clues trying to tell you a story. Your understanding of the message will help you to operate your stove more effectively.

Amount. If you burn 50 pounds of coal each day in mid-winter, a coal with an ash content of 10% will produce five pounds of ash for each day. If the amount of ash you remove seems excessive, you may be burning coal with an ash content higher than 10%. It will burn, but will require more tending and produce fewer BTU's per ton than a coal with lower ash content. Try to buy coal with an ash content of 12% or less.

Consistency. Chunks of unburned coal indicate that the fuel is undersized for your stove. They may also mean that your shakedown technique is too frequent or too vigorous.

Wood Ash

Woodburners will be primarily concerned with the consistency of the ash. Lumps of unburned charcoal represent BTU's that should be helping to heat the house, and are the result of insufficient oxygen. In fact, the traditional method of making charcoal is under oxygen-starved conditions. A hardwood fire is smothered in an airtight kiln and allowed to smolder. The volatiles are driven off and the charcoal, or fixed carbon, remains. Fixed carbon may represent up to 40% of the available BTU's.

A Pollution Perspective

Pollution in its current sense, is a relatively new addition to the world's vocabulary. Twenty years ago stoves did not pollute, they emitted smoke. Automobiles did not pollute, they emitted exhaust. And factories did not pollute, they discharged effluents. The concern about pollution diminishes as one ventures further into the American past. Human activities were more spread out 150 years ago, and there was little concern about contamination of the environment. As cities and eventually rural areas became more densely populated, it became harder for us to ignore the noxious elements given off by our activities. Pollution is, and has always been, a function of time, space, and density.

In recent years, the element of time has become critical in understanding the effect of pollution on the environment. Unlike the effluents of the 19th century, our modern toxic chemicals that are discharged into the environment are not readily altered by the action of the sun. Some of these chemicals have the ability to alter the characteristics of our unborn for generations to come.

The factor of time is well understood if one thinks of it as how long it takes nature to heal the wounds created by man. For example, one of the

A Supertanker

THE ULCC'S (ULTRA LARGE CRUDE CARRIERS, VESSELS OVER 400,000 TONS) ARE THE BIGGEST SHIPS EVER BUILT.

THEIR CARGO HAS THE POTENTIAL, IF SPILLED, TO CREATE THE BIGGEST OCEAN MESS EVER MADE.

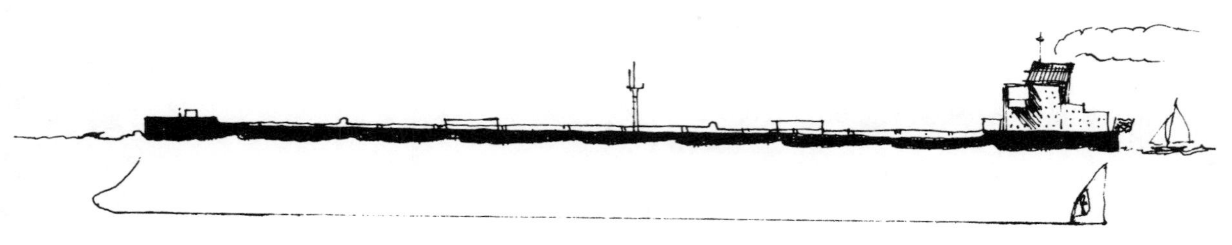

by-products in the smoke that rises up your flue when you operate your stove is carbon monoxide. In large concentrations, carbon monoxide is poisonous and can cause death. By an ironic quirk, man's blood would rather absorb carbon monoxide from the atmosphere than oxygen. However, as the carbon monoxide leaves your chimney it is rapidly oxidized in sunlight to the more benevolent carbon dioxide that is needed for vegetation to survive. This process can take less than 24 hours.

Consider, in contrast, the home that uses electric heat produced by a nuclear generating plant. Although the house itself is contributing no pollution to the environment, the nuclear plant is creating mutagenic plutonium as an atomic waste. The health implications of plutonium may exist for thousands of years.

A reliance on energy forms that carry the potential of irreversible damages is a form of deficit spending in which the interest costs of the borrowing will have to be borne by future generations. The situation is similar to the survival struggle of the old Vermonter, who as he approached old age put aside more wood each year than he consumed. This was his "retirement fund," a bulging woodshed that would be close by when he was less able to work. However, an exceptionally severe winter found the grim scene of snow still on the ground and an empty woodshed, because the old man had outlived the retirement fund.

By depleting our natural resources and risking long-term environmental destruction, we may be consuming our "retirement fund" at an uncomfortably rapid rate.

Acid Rain

In 1982, it was confirmed that six ponds in Vermont had been "killed" by acid rain. Basically, the pH level, or acid/alkaline balance, had tilted so far in the acid direction that fish and some forms of plant life could no longer be supported.

Acid rain is an atmospheric pollution problem attributed to the large scale burning of bituminous coal in industrial and power generating plants, many of them located in the states west of Vermont.

The residential use of anthracite coal for home heating is not a major factor in the acid rain problem, since anthracite is a fairly clean-burning fuel. Bituminous, however, is not recommended for home use.

The Impact of Wood Smoke

It is necessary to examine the concerns about wood smoke before a complete personal perspective on the pollution problem can be established. Recent publicity on wood smoke has pointed to ingredients that are believed to be carcinogenic. Called polycyclic organic compounds, or more commonly POM's, this specific category of matter is still poorly understood. POM's fall into the same carcinogen class as the compounds found in tobacco smoke. And after billions of dollars have been spent by the tobacco industry to determine what levels and concentrations are necessary for a carcinoma to be induced, the answer is still unclear.

A second complicating factor in the search for the answer is the methods of measurement. Although POM's are measurable in wood smoke under laboratory conditions, they are very difficult to measure in atmospheric conditions, even in communities where wood burning appliances are heavily used. The reason for the difficulty in measuring is the infinitesimally small concentrations.

Little understanding can be attained by looking at the question historically, either. Many communities in Vermont have populations that have remained stable, or in some cases declined, since 1850. All home heating was done by wood at that time, but there is no evidence that significant carcinoma-related health problems existed then.

At this time, it can be said that little scientific progress has occurred that could legitimately indict wood smoke as a health problem. Despite the lack of any incriminating evidence, a number of private enterprises within the woodstove industry are in the process of developing solid fuel appliances whose emissions are significantly reduced. The new generation of efficient, clean-burning stoves can be used in conjunction with sensitive and appropriate forest management techniques to continue to be part of the long term energy solution.

Stoves of the Future

To understand the designs and concepts that have begun to appear in the "next generation" of solid fuel appliances, one must first consider the factors that have motivated stove designers to digress from steadfast, tried-and-true approaches.

There are two main factors that provide the driving force behind new stove design. First, there is the matter of solid fuel economics. Several years ago during the initial resurgence of solid fuel use as an alternate means of heating a home, the consumer was preoccupied with the vast amounts of fuel oil he was saving by burning a few cords of wood as supplementary heat. All the stove manufacturer had to do was to keep up with orders. Even stoves that were designed around existing, but decades old, technology seemed quite adequate.

Most of those new woodburners have become seasoned veterans. They have begun to think about solid fuels as the norm instead of the novelty. This has led to a demand for improvement in appliance performance, with the obvious implication being less wood for the same heating performance. Less wood used means less work and fewer dollars spent. The demand for solid fuel heating economies, then, provides the first motivation for the stove designer.

The second impetus for improvement is the growing appreciation for environmental purity. Although it has been difficult to date to document the impact of solid fuel pollution, most of us want our air quality to remain as natural as possible. Stoveowners, after all, often begin burning wood or coal as a result of their concern to preserve natural resources.

Both of these forces point to the same design challenge; that is, improve the combustion and overall efficiency of stoves. Do this, and the concerns about economy and the environment are both satisfied.

How can stove efficiencies be improved? One method is to bring technology from other areas into the area of solid fuel combustion. One promising concept is the catalytic combustor, and much of the knowledge of combustors can be credited to work done with catalytic devices in the automobile industry.

The idea is basically sound. A catalyst works by altering a chemical reaction without actually entering into the reaction itself. In the case of woodstoves, the chemical reaction is combustion and the altering action is the lowering of the temperature required for a wood smoke/air mixture to ignite. While a conventional woodstove requires temperatures of 1100°F or 1200°F to sustain ignition of volatile gas/air mixtures to promote "secondary combustion," a properly designed catalytic stove can accomplish the same thing at temperatures hundreds of degrees lower. Consumers often operate their conventional stoves at burn rates and temperatures too low to sustain normal "thermal" secondary combustion but still hot enough to promote catalytic combustion. Catalytic combustors also get extremely hot during operation. This can provide an excellent opportunity for thermally-induced combustion in addition to catalyst-enhanced combustion. If volatile materials that might have escaped are burned, the emissions will be lower, and the efficiency (if the extra heat generated can be captured) is increased.

There can be drawbacks to catalytic combustors. The catalysts currently being used are suitable for wood combustion only. They can be ruined by materials such as sulfur that occur in other fuels like coal. Therefore, care must be taken not to burn *anything* but wood with them. Things like plastic, fire starters, or even colored printed matter all contain materials that can render the catalyst useless. Also, questions remain about the long-term durability of combustors under even normal use. All these things considered, the most important factor that determines the effectiveness of a combustor is the design into which it is incorporated. As more is understood about the operating characteristics and limitations of combustors in wood stoves and as materials improve, combustors should become a viable and legitimate design option as a step toward the goals of increased stove efficiency and reduction of emissions to the atmosphere.

Another new design concept has taken a completely different approach to improving stove performance. Rather than adding an extra device to the system (catalyst), designers have concentrated on improving combustion using well-established thermodynamic principles. The key to the success of these designs is in understanding as much as possible about the combustion principles and optimizing designs around them. Rather than treating combustion as a mixture of the various combustion processes mentioned earlier in this chapter, designers have tried to isolate these processes within the stove.

One design that recently entered the market isolated the combustion of wood fuel into two distinct processes; burning of volatiles during pyrolysis states, and burning of the remaining charcoal. This is a "batch concept" that works best with small fuel loads. Briefly, it works like this:

After kindling the stove (note that this is strictly a woodburning device), additional wood is added and burned at a high firing rate until the stove has reached a high temperature. Once the stove reaches the proper temperature, the pri-

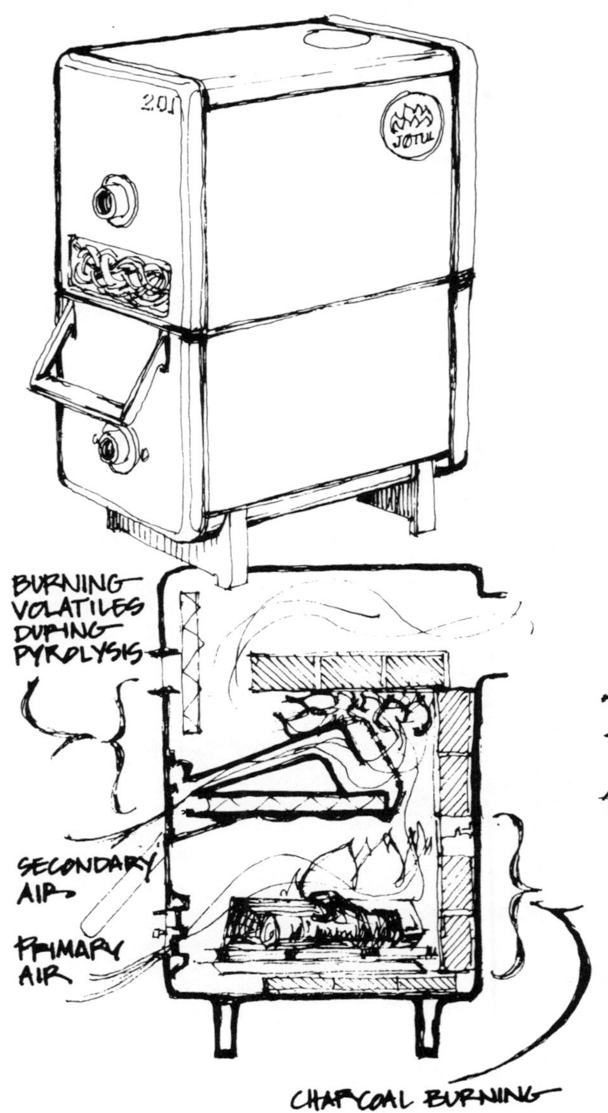

The Jøtul 201 Turbo
A Two Stage Batch Burner

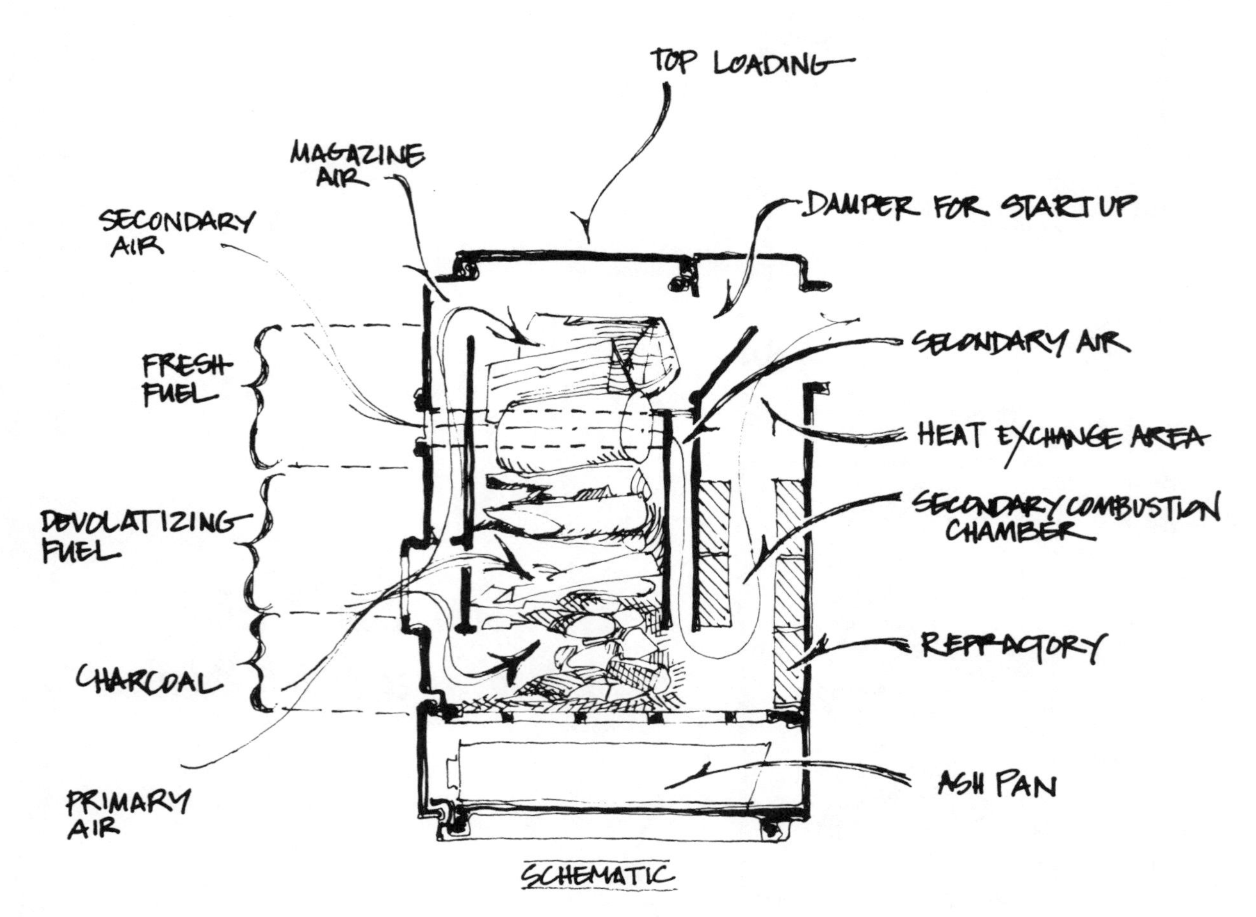

A Two Stage Continuous Burner

mary air is reduced. This allows pyrolysis or gasification of the wood load. The primary combustion zone is surrounded with insulating refractory firebrick to help maintain high primary zone temperatures and steady wood devolitization. As these wood gases exit the primary zone, they are mixed with the proper quantity of highly preheated secondary air. In a well-insulated chamber, this causes the gases to ignite and burn. The secondary air is preheated in a heat exchanger that is exposed to the combusted gases before they exit the stove.

Once the volatile materials have been driven from the wood, the secondary air supply is closed, and the remaining charcoal combusts on primary air only. (Secondary air isn't required for good charcoal combustion.) This batch technique extracts heat from both burning phases without compromising either.

Based on similar principles are other pending designs which have tried to eliminate the small load/batch-burning inconvenience. These include much larger fuel magazines and a nearly constant evolution of the pyrolysis gases for very long periods of time. Secondary combustion beyond the primary level is treated in a similar manner. These devices with larger fuel magazines are also suitable for fuels other than wood logs. Nearly any of the fuels classified as biomass fuels can be handled in this system. A successful design of this type would have tremendous fuel flexibility and would have significant impact on both economic and emissions factors of solid fuel burning.

Heat storage units, although not space heaters, also offer good overall efficiencies and very low emissions. The principle, used in a simpler form for years in the tried-and-true design of the Russian Fireplace, involves generating an extremely high temperature/high turbulence environment for wood combustion by forcing air through the fuel zone (with a blower) and by burning the wood rapidly. The tremendous quantities of energy released in short periods of time are then captured "downstream" in a water storage system. There are some drawbacks though; the systems are expensive and large, and the potential for heat loss from the water storage must be considered.

A final concept worth mentioning, called "co-generation," also draws on technology being pursued in other energy fields. In this system, a solid fuel is gasified to produce a highly burnable gas. This gas can be converted directly to heat by burning or can be used to power an engine to produce electricity. This concept provides for the total energy needs of the user and also provides for the most efficient use of solid fuel. Emissions would be negligible if not non-existent as is typical with gas combustion. This may indeed be the "Stove of the Future."

Future Fuels

In his resourcefulness man has burned any variety of things to keep warm. Settlers of the American West, for instance, burned buffalo chips for lack of wood. During the Depression years, many families in rural areas used dried corn cobs in place of wood or coal.

When the energy shortage of the 1970's made the switch to coal and wood a reality, many

Do You Have What It Takes?

By this time the reader should have a good idea of what is involved in Stove School. Perhaps you would like to chuck your career as nuclear physicist or brain surgeon and take up the simple life of a stove expert in the north woods. On a busy Fall day, a Customer Relations representative at Vermont Castings may deal with more than one hundred customers and potential stove owners from all over America.

The needs of customers run the gamut from ridiculous to sublime. An engineer from New Jersey called to ask us the weight of a ton of coal. A lady from Massachusetts requested instructions for converting her stove into an aquarium. A Californian ordered two stoves to use as end tables for his couch.

Some calls approach the limits of credulity. A man called once to ask instructions on disassembling his stove. When we offered to send them, he sheepishly asked if we would mind reading them over the phone. It seems his five-year-old daughter had been playing inside the stove and was now inextricably stuck. A customer from Maine asked if we could send him a new thermostat at no charge. He had arrived home, the newly purchased stove in the back of his pick-up, and rushed in to tell his wife. Unfortunately, he neglected to put on his parking brake. The truck rolled down the driveway, through a small retaining wall, and down a cliff. The truck was demolished; the stove, save the thermostat, was intact.

If you think you have what it takes to handle calls like these, and if you are willing to be overworked and underpaid, then give us a call. Just don't forget, rookies get stovestoking duty in the showroom.

people supplemented their fuel supply with newspaper logs. The logs could be rolled by hand at home in the living room and would produce a hot but short fire. The main objection to this unique fuel was the labor intensive procedures of preparing them and the unpredictability of their burning performance.

The alternative fuel with the most promising future is pelletized wood. A high quality solid fuel can be produced from the wood wastes that are currently discarded by the lumber and paper industries. Using a process similar to that used in the manufacture of pelletized animal feeds, convenient small pellets are produced. These clean, dry pellets may be stored safely and create a very hot flame when burned. The intense heat is one of the reasons that many parlor stove manufacturers do not recommend burning wood pellets in their units. Although they may be burned successfully in some coal stoves, the pellets can also be highly volatile, and the fire can be difficult to control in a parlor wood stove.

UP TO 999,000 BTU/HR, ON PELLETS, CHIPS, SCRAP — WITH STOKER, SKID, HEAT EXCHANGER, BLOWER...

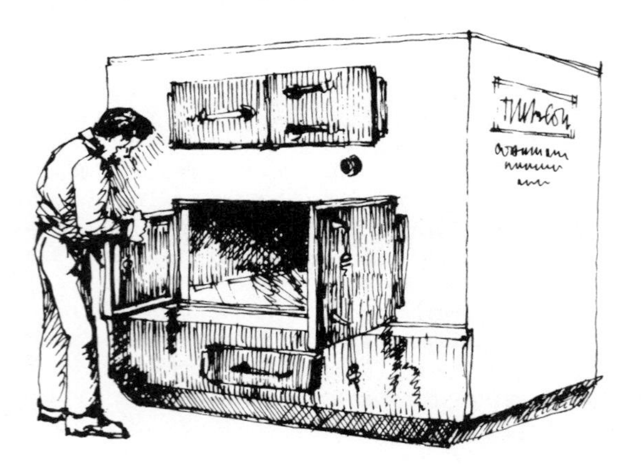

Industrial Furnace

Best results with wood pellets are achieved when they are used in large industrial heaters or in domestic central heating systems. Either an automatic auger-feed system or some sort of hopper-feed system is usually employed when burning wood pellets.

In addition to wood, pellets are also being produced from cardboard and garbage.

A more readily available source of alternative fuel is wood chips. Unrelated to the buffalo chips mentioned earlier, they burn well in certain stoves, central heating systems and gasification systems. However, they are not recommended for use by most parlor stove manufacturers.

Sawdust is also readily available and is quite inexpensive. It can be burned in special stoves provided that a supply can be found with a moisture content of less than 60%. If the sawdust has a moisture content higher than 60%, most of the heat from the fire is used to dry out the fuel. At least three Pacific Coast firms are manufacturing briquets from sawdust. These briquets are clean and easy to handle. They produce very little ash, no creosote, and little smoke. If these are available in your area, you may be able to burn them in your parlor stove. Check with your manufacturer first.

Doctor of Stoves

Physician, heal thyself. By winter the stove owner is a graduate of Stove School, and his "book learnin'" is being put to the test daily in a non-stop effort to fend off the chill of winter, not to mention the oil delivery truck. Man, however, does not count among his innate skills the ability to burn an efficient, airtight wood or coal stove. No matter how thorough the advance preparation, no matter how intelligent the operator, when the temperature plummets and the new stove faces its first severe challenge, problems are inevitable.

We have tried in this book to distill our experiences with 100,000 tutors out in Stoveland to provide the basic information that a stoveowner needs to coax the maximum performance from his unit. Our success rate would be the envy of any medical doctor or diagnostician, so by most standards we are entitled to think of ourselves as experts. In truth, however, our success rate in solving problems is deceptive. More often than not, the solution to a problem is not provided by the doctor but by the patient.

The following illnesses account for nearly all of the problem calls we receive: low heat, short burn time, creosote, smoking, odor, or any combination thereof. Variable causes can be related to the installation, operation techniques, expectations of the operator, or even the weather. Thus, a seemingly simple problem may be the result of an infinite combination of variables.

We have adopted a Socratic method of assisting customers, emulating the technique of the Greek philosopher who taught his students by answering their questions with questions of his own. The trick is in knowing the right questions to ask, and as any veteran of the Vermont Castings Customer Relations staff will attest, the most important question is always the one left unasked. For example, a man from Long Island with an acute creosote problem patiently exhausted the resources of our most experienced experts. Everything seemed right—good installation, the right fuel—and yet every morning would see new rivulets of gooey black fluid appearing. The customer himself was cooperative and seemingly intelligent, and yet his stove resisted every proffered solution. Not until we actually saw his stove did the cause of his woes become obvious—there was a five-ounce fishing sinker attached to his air inlet damper which kept it permanently shut. A frustrated Customer Relations representative, who had spent many long hours educating the customer to the nuances of stove

operation, called back to ask, "Why?" The man replied that his wife had devised the sinker to eliminate the slight clinking noise which occurred whenever the damper closed. The Customer Relations person could contain himself no longer and blurted, "Why didn't you tell us that you had a five-ounce lead weight on your air inlet damper." The response was as simple as it was frustrating: "You never asked."

To save time, we have formalized our checklist approach and include it here, noting why each question is asked. We send this out whenever a problem is complex enough to defy easy solution. Only a fraction are returned, a fact that perplexed us when we first began. As we followed up to learn the reasons for this lack of responsiveness, however, we were pleasantly surprised to learn that many customers had solved their own problems. In the process of answering our questions, they had answered their own. At some time in your wood or coal burning career you will encounter problems. By reviewing this checklist you might help yourself to an easy solution.

Some of the many factors that affect a stove's heat output and general performance are obvious, such as type, moisture content, size, and shape of firewood. Others are more difficult to define and require detailed information regarding the installation and operation of the stove. First, you must define the problem. Are your expectations reasonable, or are you trying to heat a football stadium with a stove designed to heat a four-room house? Ask yourself some basic questions.

When Did the Problem Start? Have you operated your stove long enough to give it a fair chance? Have you read the operation manual to make sure you are not doing something wrong that is obvious? If the change in performance is sudden, it may indicate a change in the weather, installation or wood supply, and not a stove problem.

Has the Problem Gotten Worse? If your stove has been working properly but performance has gradually deteriorated, the likely culprit is creosote build-up. Don't blame the stove, blame yourself.

Is the Problem Constant? An intermittent problem should tell you to look at the changeable variables, such as weather or wood supply. A problem that defies changes in operation is indicative of stove or installation faults.

The Chimney

Draft depends on three factors: the size and configuration of the chimney and flue pipe, the height of the chimney, and the temperatures of both the gases within the chimney and the outside air. Although certain types of chimneys are preferable to others, none are immune from problems. The sources of problems can range from leaky clean-out doors to creosote-clogged bird screens, so a useful first step is to diagram your installation before answering the following questions.

Masonry or Prefabricated? Masonry makes an ideal chimney material but is not foolproof. The potential problems that can affect a masonry

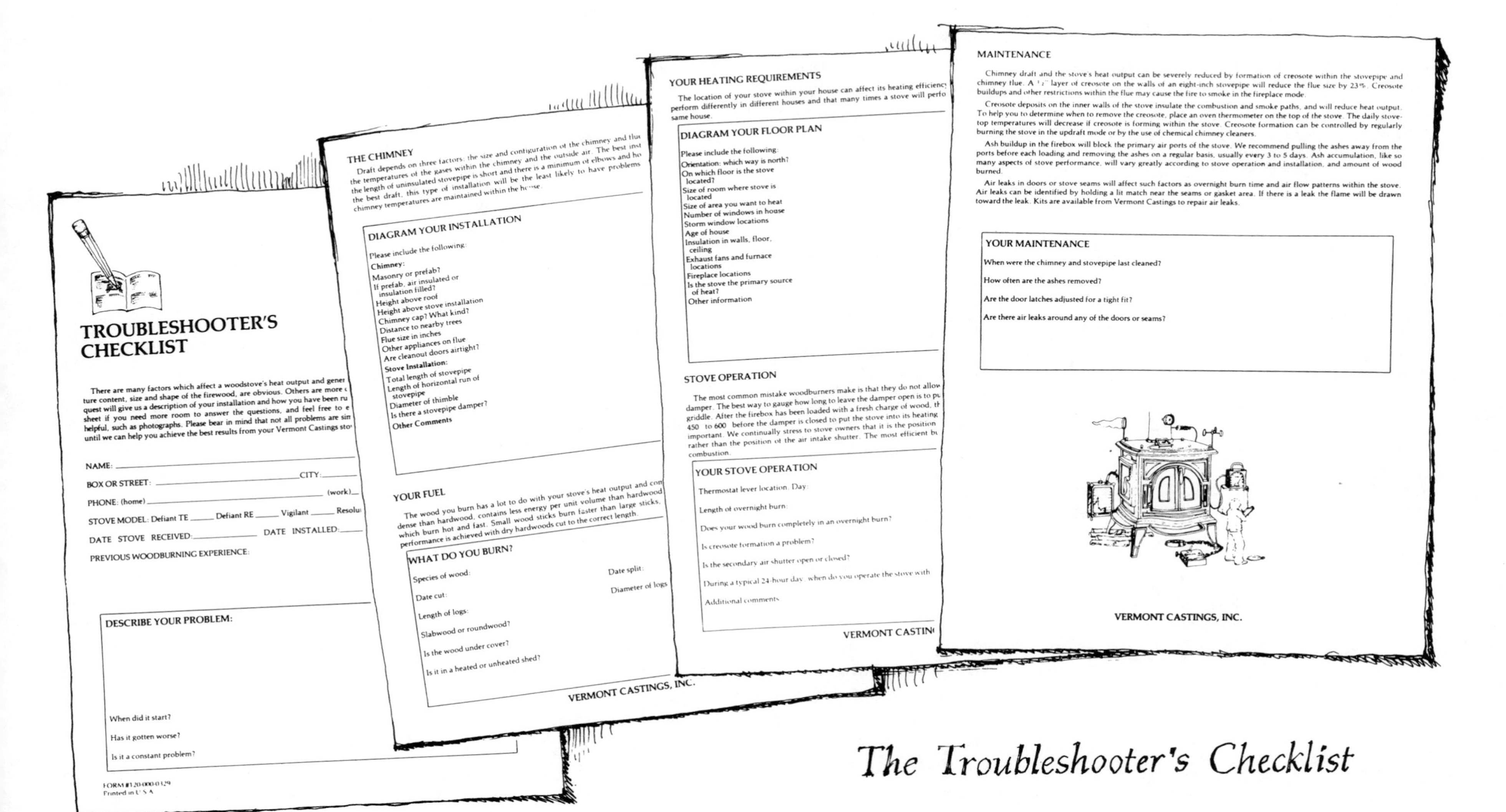

TROUBLESHOOTER'S CHECKLIST

There are many factors which affect a woodstove's heat output and gener ture content, size and shape of the firewood, are obvious. Others are more c quest will give us a description of your installation and how you have been ru sheet if you need more room to answer the questions, and feel free to e helpful, such as photographs. Please bear in mind that not all problems are sim until we can help you achieve the best results from your Vermont Castings sto

NAME:

BOX OR STREET: CITY:

PHONE: (home) (work)

STOVE MODEL: Defiant TE ___ Defiant RE ___ Vigilant ___ Resolu

DATE STOVE RECEIVED: ___ DATE INSTALLED:

PREVIOUS WOODBURNING EXPERIENCE:

DESCRIBE YOUR PROBLEM:

When did it start?

Has it gotten worse?

Is it a constant problem?

FORM #120-000-0129
Printed in U.S.A.

THE CHIMNEY

Draft depends on three factors: the size and configuration of the chimney and flue the temperatures of the gases within the chimney and the outside air. The best inst the length of uninsulated stovepipe is short and there is a minimum of elbows and ho the best draft, this type of installation will be the least likely to have problems chimney temperatures are maintained within the house.

DIAGRAM YOUR INSTALLATION

Please include the following:

Chimney:
Masonry or prefab?
If prefab, air insulated or insulation filled?
Height above roof
Height above stove installation
Chimney cap? What kind?
Distance to nearby trees
Flue size in inches
Other appliances on flue
Are cleanout doors airtight?

Stove Installation:
Total length of stovepipe
Length of horizontal run of stovepipe
Diameter of thimble
Is there a stovepipe damper?

Other Comments

YOUR FUEL

The wood you burn has a lot to do with your stove's heat output and con dense than hardwood, contains less energy per unit volume than hardwood which burn hot and fast. Small wood sticks burn faster than large sticks. performance is achieved with dry hardwoods cut to the correct length.

WHAT DO YOU BURN?

Species of wood:
Date split:
Date cut:
Diameter of logs
Length of logs:
Slabwood or roundwood?
Is the wood under cover?
Is it in a heated or unheated shed?

VERMONT CASTINGS, INC.

YOUR HEATING REQUIREMENTS

The location of your stove within your house can affect its heating efficiency perform differently in different houses and that many times a stove will perfo same house.

DIAGRAM YOUR FLOOR PLAN

Please include the following:
Orientation: which way is north?
On which floor is the stove located?
Size of room where stove is located
Size of area you want to heat
Number of windows in house
Storm window locations
Age of house
Insulation in walls, floor, ceiling
Exhaust fans and furnace locations
Fireplace locations
Is the stove the primary source of heat?
Other information

STOVE OPERATION

The most common mistake woodburners make is that they do not allow damper. The best way to gauge how long to leave the damper open is to pu griddle. After the firebox has been loaded with a fresh charge of wood, th 450 to 600 before the damper is closed to put the stove into its heating important. We continually stress to stove owners that it is the position rather than the position of the air intake shutter. The most efficient bu combustion.

YOUR STOVE OPERATION

Thermostat lever location: Day:
Length of overnight burn:
Does your wood burn completely in an overnight burn?
Is creosote formation a problem?
Is the secondary air shutter open or closed?
During a typical 24-hour day, when do you operate the stove with
Additional comments

VERMONT CASTIN

MAINTENANCE

Chimney draft and the stove's heat output can be severely reduced by formation of creosote within the stovepipe and chimney flue. A ½" layer of creosote on the walls of an eight-inch stovepipe will reduce the flue size by 23%. Creosote buildups and other restrictions within the flue may cause the fire to smoke in the fireplace mode.

Creosote deposits on the inner walls of the stove insulate the combustion and smoke paths, and will reduce heat output. To help you to determine when to remove the creosote, place an oven thermometer on the top of the stove. The daily stove-top temperatures will decrease if creosote is forming within the stove. Creosote formation can be controlled by regularly burning the stove in the updraft mode or by the use of chemical chimney cleaners.

Ash buildup in the firebox will block the primary air ports of the stove. We recommend pulling the ashes away from the ports before each loading and removing the ashes on a regular basis, usually every 3 to 5 days. Ash accumulation, like so many aspects of stove performance, will vary greatly according to stove operation and installation, and amount of wood burned.

Air leaks in doors or stove seams will affect such factors as overnight burn time and air flow patterns within the stove. Air leaks can be identified by holding a lit match near the seams or gasket area. If there is a leak the flame will be drawn toward the leak. Kits are available from Vermont Castings to repair air leaks.

YOUR MAINTENANCE

When were the chimney and stovepipe last cleaned?

How often are the ashes removed?

Are the door latches adjusted for a tight fit?

Are there air leaks around any of the doors or seams?

VERMONT CASTINGS, INC.

The Troubleshooter's Checklist

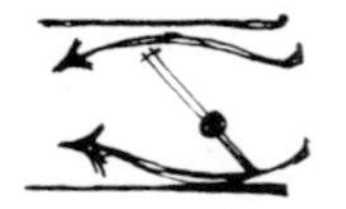

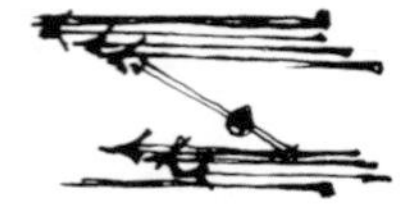

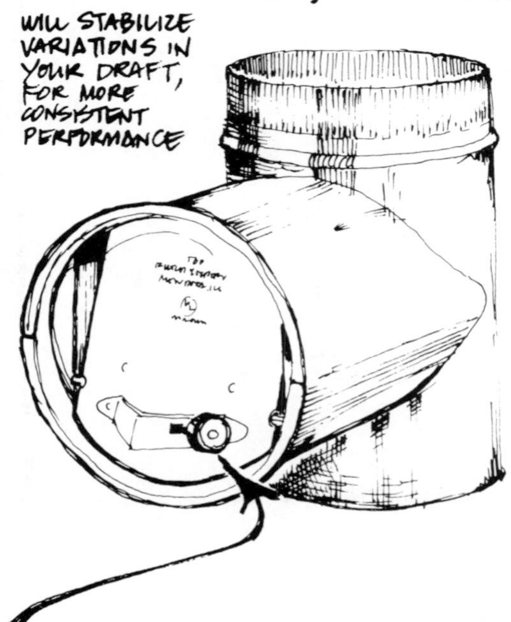

chimney constitute a completely different set of variables than those which can plague prefabricated units and vice versa.

Prefab—Air Insulated or Insulation Filled? The various types of prefabricated chimneys have entirely different properties for insulating flue gases. Knowing which type you have is a key starting point in determining if your chimney is well-suited to the needs of your stove.

Height Above Roof? Most building and fire codes state that a chimney has to be a minimum of two feet above the peak of the house if within ten feet of the peak. If the chimney is farther than ten feet away, it must be more than three feet above a horizontal measurement of ten feet from chimney to roof line. These minimum heights are for safety purposes and also to avoid turbulence generated by air currents over the roof. It is interesting to note that older homes in Vermont (when wood was the primary fuel) have chimneys five to seven feet above the peak. Most homes built between the turn of the century and 1940 have chimneys about three to five feet above the peak, and modern homes have chimneys about two feet above the peak, an indication both of inflation and our decreased reliance on our hearths. Perhaps we could learn from our forefathers.

Height Above Stove Installation? If a chimney is too short, this can cause a low draft problem. Conversely, a chimney that is too high can result in the opposite problem—excessive draft. Generally speaking, a minimum of fifteen feet of chimney is necessary to ensure adequate draft.

Chimney Cap—What Kind? Chimney caps are designed to serve a variety of functions, but primarily to keep the flue dry. Creosote from woodburning and fly ash from coalburning form strong corrosive acids when mixed with rain water. Other chimney caps are intended to correct specific draft shortcomings. Too often, however, the cap can be part of the problem rather than the solution.

Distance to Nearby Trees? Tall trees too close to a chimney can cause air turbulence which affects draft

Flue Size in Inches? If the area of the flue is too small or excessively large, a draft problem is inevitable.

Other Heating Appliances on Flue? Besides representing a potential safety problem, multiple appliances on a single flue can allow too much cool air to be drawn into the chimney.

Are Cleanout Doors Airtight? If the cleanout door at the base of the chimney is not airtight, cold air will be sucked into the chimney, cooling the flue gases and reducing their buoyancy. The result? Excessive creosote and a host of operational problems.

Stove Installation—Total Length of Stovepipe? The metal stovepipe that connects the stove to the flue has entirely different insulating characteristics than masonnry or prefabricated pipe. Specifically, it allows heat to radiate rapidly; hence, too long a run will cool the gases quickly. How long a run is permissible depends on the strength of the entire system and cannot be stated categorically.

Total Length of Horizontal Run of Stovepipe? No more than a four-foot run of horizontal pipe or not more than two elbows in any system is ever advisable. The shortest, straightest possible run of stovepipe is always the best design.

Diameter of Thimble? The diameter of the thimble should be the same as the diameter of the stovepipe to permit the hot gases to transit from the stove to the chimney in a smooth, unrestricted manner. A reduction at this point will cause excessive turbulence which will be evidenced by smoke coming into the room anytime you open the stove doors.

Is There a Stovepipe Damper? A stovepipe damper is not necessary with most air controlled stoves, and in most cases, will hinder performance rather than help. As with all rules, there are exceptions such as in cases of excessive draft when a stovepipe damper or barometric draft stabilizer may be necessary.

Your Fuel—Wood

The wood you burn reflects directly on your stove's heat output. Softwood, being generally less dense than hardwood, contains less energy per unit volume than hardwood, and also contains resins which burn hot and fast. Small wood sticks burn faster than large sticks, dry wood faster than wet. The principles are simple enough, but in practice we have found that the "two-year-old oak" which the customer just bought from his local dealer turns out to be poplar with green leaves still attached. These questions help establish a common language for analyzing problems:

Species of Wood? Woods vary in density and therefore in BTU potential. Don't expect the same heat output and burn time from a load of willow as a load of apple.

When Was the Wood Cut? Wood should be cut and dried for a minimum six months. As with fine wine, it will improve with aging unless rot sets in.

How Long Are the Logs Being Burned? Burn time can often be related to log length. A stove designed to take eighteen-inch logs in which you are burning twelves will be losing one third of its capacity. This means that the stove, which should hold a fire for nine hours, will die after six. Underutilization of firebox capacity is the most common cause of short burn times.

Do You Burn Slabwood or Roundwood? Slabwood and round wood have different drying characteristics. With so much of its cellular surface exposed, slabwood is like kindling and will burn hot and fast. Since it releases so many volatiles it will sometimes cause the explosive puffs known as "backpuffing."

Is the Wood Under Cover? Covered wood dries more quickly and stays drier than wood exposed to the elements.

Is It a Heated or Unheated Shed? Wood in a heated building will continue drying once the heating season starts and is preferable.

When Was the Wood Split? The earlier the better. Unsplit wood has a higher moisture content than split.

What Is the Diameter of the Log? The heftier the log size, the more slowly it will burn. The immediate heat output will be lower as well.

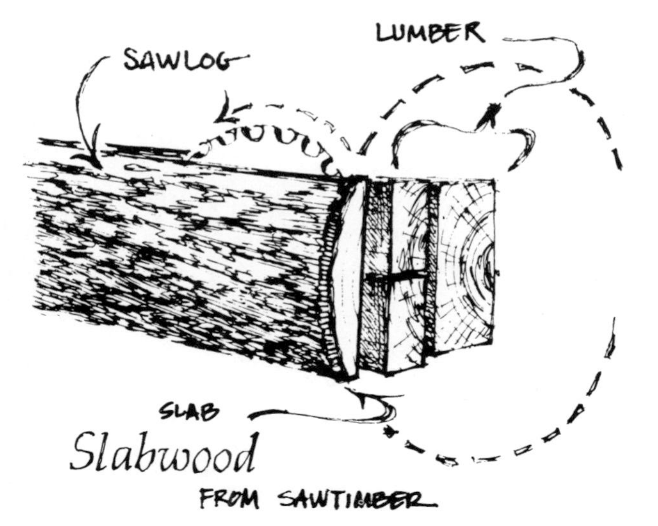

Your Fuel—Coal

Although there are many similarities in wood- and coalburning when it comes to fuel-related problems, a different troubleshooting approach is required. Because there are no units available to the homeowner which burn bituminous coal at an acceptable level of pollution emission, discussion here is limited to anthracite.

What Is the Ash Content? This should be available from a reputable dealer, but don't be surprised if he doesn't know what you're talking about.

What Size Coal Are You Burning? Size affects hopper adjustment in an automatic feed system. The smaller the coal the less the air space in the fuel bed. Operating techniques have to be adjusted accordingly.

How Frequently Do You Shake, Slice, and Remove Your Ashes? Unlike wood, coal burns from the bottom, requiring air to come up through the grates. Proper shaking and slicing is necessary to clear the grates of the dead ash so the fire can breathe. It is important that ashes are not allowed to build up in the pan. This can restrict the flow of air under the grates.

Your Heating Requirements

The location of a stove within the house can affect its heating efficiency. The same stove can perform differently in different houses or even in different flues within the same house.

As part of the checklist approach the customer is asked to diagram his floor plan. This section is left blank or labeled irrelevant more than any other portion of the exercise. Perhaps its purpose is not immediately obvious, but it has been our experience that more complaints are due to unrealistic expectations than any other single cause. Surprisingly, more stoves are oversized for their heating requirements than undersized. Insight on a wide variety of problems can be gained from the information sought here. The location of a stove within a room, the room within the home, and the orientation of the home all will have bearing on stove performance.

Orientation: Which Way Is North? An exterior chimney located on the north side of a building will lose heat much faster than one located on a warmer side. Loss of heat in the chimney causes a reduction of draft, a cooler burning stove, and creosote. The geographic orientation will also determine how prevailing winds or breezes may affect the chimney.

Size of Area You Want to Heat? Many people are trying to heat only one or two rooms of the home, but, because of the layout, are dissipating BTU's into adjacent areas.

Size of Room Where Stove Is Located? A stove in a small room with few exits can cause that one room to overheat, while not allowing enough heat to travel into other areas. The situation often can be improved by using registers or fans to better circulate the air. Conversely, a small stove in a very open room may not heat that room well if the available BTU's are being spread over an area beyond the stove's capacity.

Number of Windows in House and Storm Windows' Location? The number of windows and their location tells a great deal about how much heat is going outside. Storm windows slow this process and thus make heating easier.

Insulation in Walls, Floors, and Ceiling? The amount and type of insulation in the house also has a bearing on how easily it can be heated. The less insulation, the more wood will have to be burned to heat the same space. Draft problems can be a problem of a very "tight" home. Smoke cannot leave the house via the chimney faster than outside air can infiltrate the structure to replace it. Although rare, this situation can occur in mobile homes and newer homes where a barrier of plastic film is used as part of the insulation. It can be alleviated by introducing a source of outside air near the stove.

Age of House? The age of the house gives clues as to the type and amount of insulation, construction of the chimney, and the general construction of the house.

Exhaust Fans and Furnace Locations? Other appliances such as exhaust fans, fireplaces, furnaces, and other stoves can interfere with the smooth operation of the stove. Exhaust fans, for instance, can remove air from the house fast enough to slow, stop, or even reverse the flow in the chimney.

Is the Stove the Primary Source of Heat? A stove used as the primary source of heat will be the subject of greater expectations than a backup. Some people have a difficult time accepting the fact that the stove which can keep the house toasty at 20° below will by definition be oversized for taking the chill off a crisp October morning.

Other Information? Remember the unasked questions? Any information, regarding solar gain, other heating sources, hills, large bodies of water, prevailing winds, or any natural or manmade phenomenon will help.

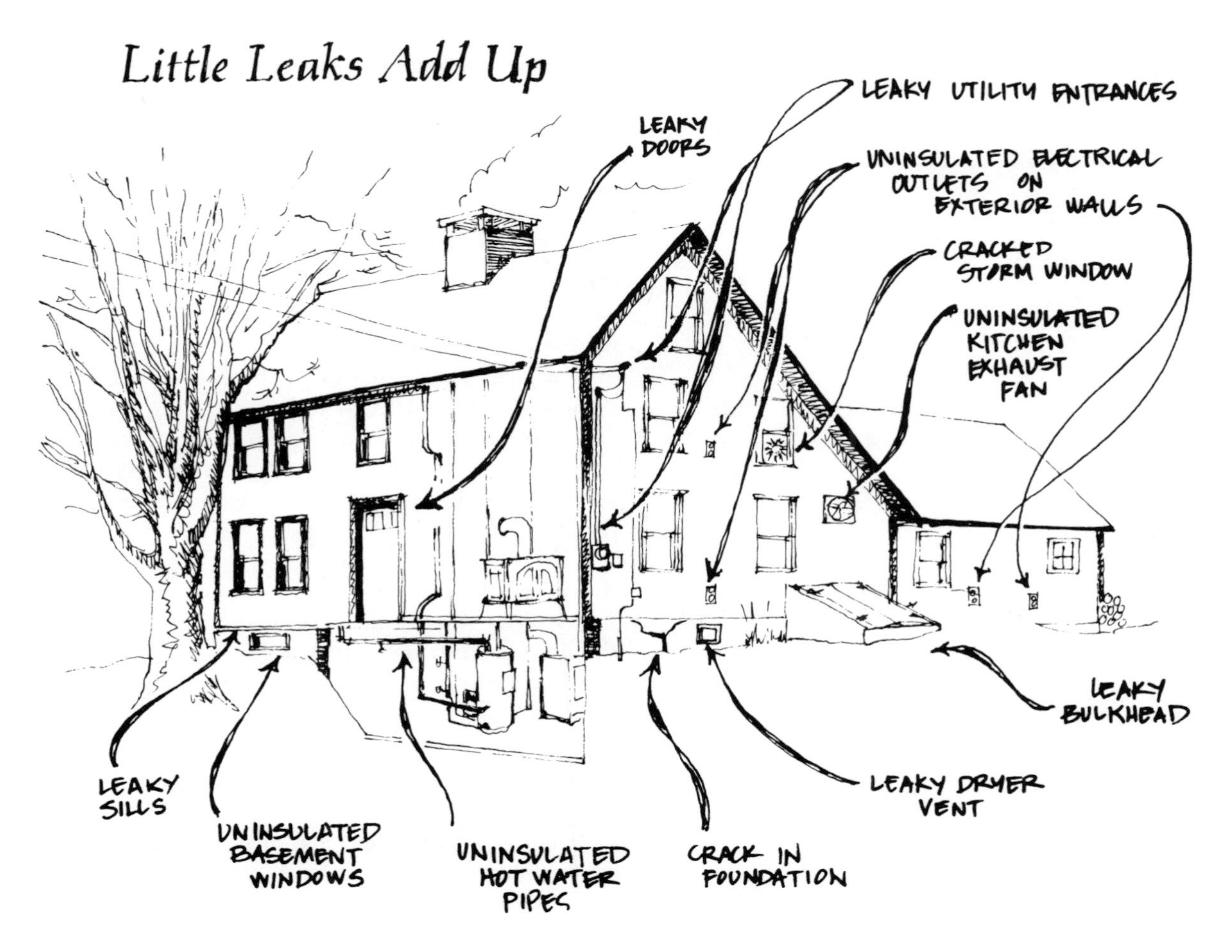

Stove Operation

The majority of stove problems are operation-related. The stoveowner needs to be attuned to the signals his stove sends. This section of the checklist asks the stoveowner to look at some basics of operation.

How Do You Control the Air Supply and the Resulting Heat Output of the Stove? The amount of oxygen being admitted to the fire will directly affect heat output. The hotter the stove, the better the draft.

Stoves use various methods of controlling oxygen intake. Thermostatically-controlled stoves, such as those made by Vermont Castings, present the operator with a more intricate operational challenge.

The thermostat is a bimetal coil that reacts to the heat of the stove. The initial setting is manual, and as the stove heats up, the bimetal coil expands, lowering the air-flap, restricting the oxygen, and thereby reducing the heat output. As the stove cools, the coil contracts, more oxygen is admitted and the combustion returns to the designated level.

Two very important factors for the owner of a thermostatically-controlled stove are the following: a) The relationship between the position of the lever and the opening of the air inlet flap will vary according to the specific point in the burn cycle. b) There is always a trade-off between heat output and length of burn. The same load of wood, if burned in four rather than eight hours, will obviously provide more heat.

How Do You Adjust Your Secondary Air Inlet? Some stoves have a controllable inlet for secondary air which can affect burn times and heat output. This adjustment can be important for wood burning but is of less consequence with coal.

How Long Will Your Stove Burn? Everyone has a personal definition of "burn time." We define it as the time from when the stove is loaded to capacity to when the fuel is consumed. The time will fluctuate according to the length and dryness of wood, area to be heated, and temperature outdoors. While an "overnight" burn in one household is midnight to 6 a.m., in another it can be 8 p.m. to 8 a.m.

Does Your Wood Burn Completely? The consistency of ash can tell the experienced stoveowner a great deal about how well the stove is performing. The ash should be a fine powder, dry, and easy to shovel out. Chunks of charcoal indicate problems that can derive from a variety of sources. Only in concert with the rest of the checklist information can a logical analysis be made.

Is Creosote Formation a Problem? The physical appearance of creosote can tell a story, as this substance runs the gamut from watery liquid to flaky, black chips. All wood creates some creosote, so you might as well learn something from yours.

How Do You Operate the Stove During a Typical 24-hour Day? Consider how the stove is operated over a 24-hour period. Or stated another way, what are the operator-related habits that have an effect on the stove. Does the teenage son pull wood from a different section of the woodshed than Dad? Does Mom open up the air inlets as soon as the kids leave for school? These are important facts in the analysis of your situation. Knowing them will teach you about your stove, and yourself.

Anything Else

Is there anything you can imagine that may have the slightest bearing on your problem? Are you taking wood from an exposed part of your woodshed? Has the weather been warmer than usual? Did your wife put a fishing sinker on the air inlet damper?

This is the place for the meaningless tidbits that by themselves mean nothing but may supply the missing piece of the entire picture. The strange wet spot on the tile under the stove may be creosote or it could be the result of a snow-covered dog or cat warming itself by the toasty stove.

A *troubleshooter* is a gun totin' hombre who always gets his man. No one at Vermont Castings would claim that we are successful in completely eradicating smoking, odor, and low heat from Stoveland. At least with a checklist in hand, however, we feel that we have half a chance.

Just a Chimney Fire

When the fire alarm interrupts the tranquility of downtown Randolph, and the trucks roll from the fire station, everyone is interested. Everyday business slows for a few minutes while shoppers and storekeepers peer out the windows and ask each other, "Where's the fire?" When the trucks return within an hour, the word spreads through town, "Oh, it was just a chimney fire."

Learn from the "Experts"

The wise woodburner will manage his heating activity so that emergency measures are never necessary.

A Chimney Fire to Remember

"I was visiting the family homestead on the Connecticut shore, an ancient dwelling with a large (1 foot by 2 foot), unlined, (and for about ten years) uncleaned chimney.

It was a damp, dreary morning. My Dad and I decided to start a fire to take the chill off the old house. A roaring blaze of dry kindling was soon crackling. Satisfied that all was going well, Dad adjourned to the kitchen to tend the coffee while I fetched the supply of wood.

Outside at the woodpile I saw a billowing cloud of black smoke coming from the chimney. A few seconds later, flames became visible and within seconds were shooting ten feet into the air. I ran back into the living room where I was greeted by a sound that reminded me of a freight train chasing a 747 jumbo jet.

My Dad, a small-framed gentleman in his mid-sixties, leaped into action. He seized my mother's card table to block off the fireplace front, hoping to cut off the oxygen supply and extinguish the fire. It seemed like a good idea at the time.

Dad got to within a foot of the fireplace with the table before the suction took over, pulling the table (Dad still in tow) up against the brickwork with a loud bang. Dad managed to jump back just as the table blew inward and disappeared. The jet roar came back immediately.

As it turned out, the only casualty of the incident was Mom's card table, although my father and I were dramatically impressed by the violent power of a chimney fire. You can rest assured that this is one chimney which is now cleaned every year."

Duncan Syme

For the frightened victim, it wasn't "just a chimney fire." It was one of the most terrifying experiences he had ever encountered within the safety of his own home. First, there was the noise; he had heard it compared to a freight train or a jet taking off, and although there wasn't time to contemplate the finer points of auditory discrimination, the roar implied an equal amount of force. And then there was the shooting column of flames erupting from the chimney, sending sparks and cinders shooting off to the sides to land on the roof. To the stoveowner with a fiery and uncontrollable volcano in the middle of his house, it is never "just a chimney fire."

Most woodburners have a chimney fire at some point in their heating careers. Soot from a coal fire can ignite and burn too, but it is less likely. Usually a chimney fire comes early in a woodburner's career, but not necessarily. Some people have more than one in a single year, and others have one or more each year for several years in a row. A single experience, though, is enough to motivate most of us to resume attendance at Sunday services, to invest in chimney-cleaning equipment, or to get on the local sweep's regular customer list.

Like any combustion event, a chimney fire needs oxygen, heat, and fuel in order to burn. All chimneys that serve a woodstove have a constant supply of oxygen and heat, but it is the addition of fuel that allows the fire to begin. The fuel is creosote.

Creosote

Creosote is gas vapor that has condensed into a liquid form. The liquid can continue to harden and can eventually take on the characteristics of a hard glaze or a baked-on tar. The condensation of the flue gases will begin to take place at temperatures under 300°F.

Like most stove problems, creosote is related to any one or a combination of three functions: the wood being burned, the technique of operating the stove, and the installation.

Green wood has long been believed to be the major culprit. Because of the moisture content, a fire with unseasoned wood does not burn hot. This leads to two interrelated events: The low heat level of the green wood fire fails to burn the escaping gases, and the flue remains cool, providing an ideal environment for condensation to occur.

Even the woodburner who uses only well-seasoned wood (cut and split for a year or more) is not immune from creosote. Dry wood tends to burn quickly, and can result in a shorter-than-desired burn time or can create more heat than comfort requires. The reaction to either possibility often is to reduce the supply of air to the stove. This slows the combustion rate, but at the same time creates a situation similar to that encountered when burning green wood: a smoky fire with unburned gases and a cool flue.

A creosote problem with dry wood is really more a matter of stove operation than fuel. Wood should be burned at fairly high levels to achieve maximum efficiency, and reducing the air supply to a point at which the fire smolders is self-defeating. Never completely close off the air supply of an airtight stove, unless you are having a chimney fire. That is the only occasion on which "no air" does more good than harm.

A third major factor when creosote forms is the installation. Any installation that is constructed in such a way that it cools quickly or allows the flue gases to remain for prolonged periods of time may cause problems. Outside chimneys are exposed to the cooling effects of the weather and will allow the gases to cool quickly. The stove must be burned hotter when an outside chimney is used, and the operator should be more generous with the amount of heat that is lost up the flue. Outside chimneys may also be boxed in and insulated to keep them warmer. Some prefabricated chimneys such as the triple-wall metal type can cool quickly and encourage condensation. Extra long runs of stovepipe within the room will rob heat from the gases, allowing them to form as creosote before they can leave the system. Another stove that is vented into the same flue from below or an open cleanout door may allow cool air to continually sweep up the chimney.

The obvious answer to a creosote problem is to burn the stove hotter. Without a stove thermometer, this advice is difficult to follow. A magnetic stove thermometer located a foot from the stove on the stovepipe will give you a constant indicator of chimney temperature. Keep in mind that the flue temperatures should stay above 300°F, and operate the stove accordingly. A drop in the temperature during the last stage of the burn cycle need not be a matter of concern: Fewer gases are being driven off at this point, and combustion is more complete.

What To Do

A chimney fire is a rigorous test of your stove and your installation. If you have a well-made stove, have observed all clearance requirements, and are using a sturdy chimney in good repair, your main concern will be sparks landing on the roof. If you have compromised on any of the above, monitor the entire situation until the fire department arrives to supervise. In any event, follow these rules:

1. Close off all air to the stove.
2. Call the fire department.
3. Take the family outside.

Never throw water in or on either the chimney or the stove. Have the chimney inspected for cracks or other damage before using it again. And finally, remind yourself of the cliche that "an ounce of prevention is worth a pound of cure."

The Muse of Stoveland

The mythology of Greece relied on imaginary figures called "muses" to inspire grace and creativity in their literature and music. The contemporary stovetender also draws inspiration from the comfort and security of his warming stove.

The winter months provide a brief interlude of laziness between the first snowfall of autumn and the premature optimism of the Spring Fling. This is a time when the woodburner can sit by the fire and do nothing . . . nothing but contemplate the wealth of his fuel supply and fantasize about things that may have nothing to do with the tedium of everyday living.

The Flying Steam Engine

The fascination with the fundamental usefulness of solid fuel energy is timeless. It has long been used for heating and cooking, and has provided energy for transportation from time to time. The best known application is the steam locomotive, a concept that lived for 100 years or more. It has been used in other modes of transportation as well. As early as the 1840's inventors had plans on paper for airborne "steam carriages."

One steam-powered device in the latter part of the 19th century consisted of a large, heavily-riveted boiler. The sturdy construction was necessary as a precaution to minimize the chance of explosion and serious injury to the pilot and spectators within the immediate vicinity. On top of the fabulous engine was mounted a cast iron auger, the conical point of which was aimed directly at the heaven it hoped to penetrate. The large iron auger alone must have weighed three tons, and the entire contraption was no doubt horrendously out of balance. Even a mighty head of steam, with smoke billowing from the stack and jets of vapor hissing from every valve and joint, was not enough to lift this majestic craft off the ground. The inventor surmised correctly that his problem was gravity, and less correctly that gravity must be a bit like suction; if he could just overcome the initial grip that gravity made on his gargantuan vehicle, all would be well.

So he fitted a steam-powered cylinder on each corner of his vehicle to provide the needed thrust. The idea worked, at least a little bit. Once fired up, the invention began to leap about the meadow like a mammoth quadruped pogo stick, trying desperately to free itself from the earth's fetters. The mechanical stress proved too great for the unfortunate engine, and it broke up, and collapsed in a sighing heap. One can almost imagine the soul of the magnificent device rising amidst the last wisp of steam and smell of hot oil.

Hopefully, the fascination with the wood- or coal-powered engine will continue into the future. Let's take a look.

The Amazing Ossum

Welcome to the year 2100. Man has long since depleted the earth's oil and gas reserves, and nuclear power has fallen out of favor since the major facility failure. Man once again relies on the sun and on wood. Wood is seldom cut from independent tracts of natural forest growth anymore, but rather from hybrid plots where eight-inch fuel trees can be grown in five year's time.

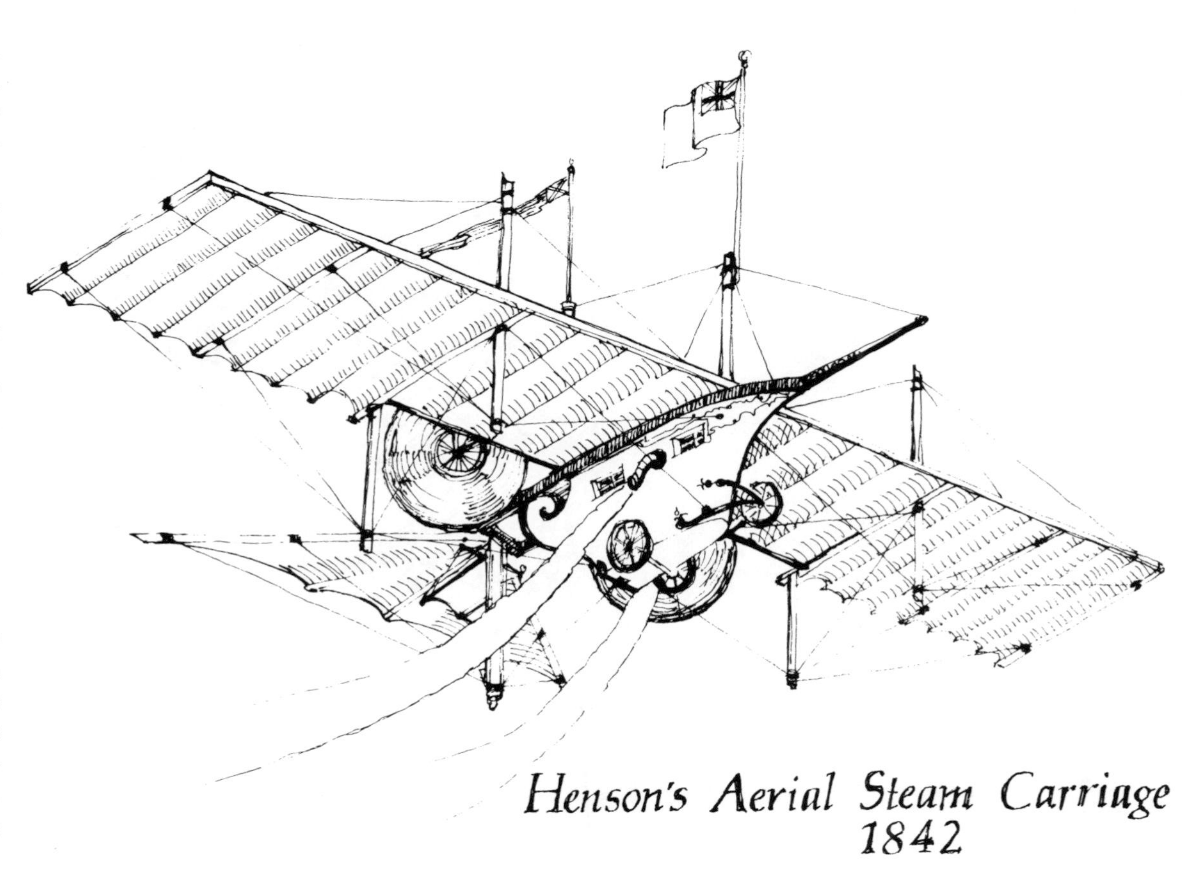

Henson's Aerial Steam Carriage
1842

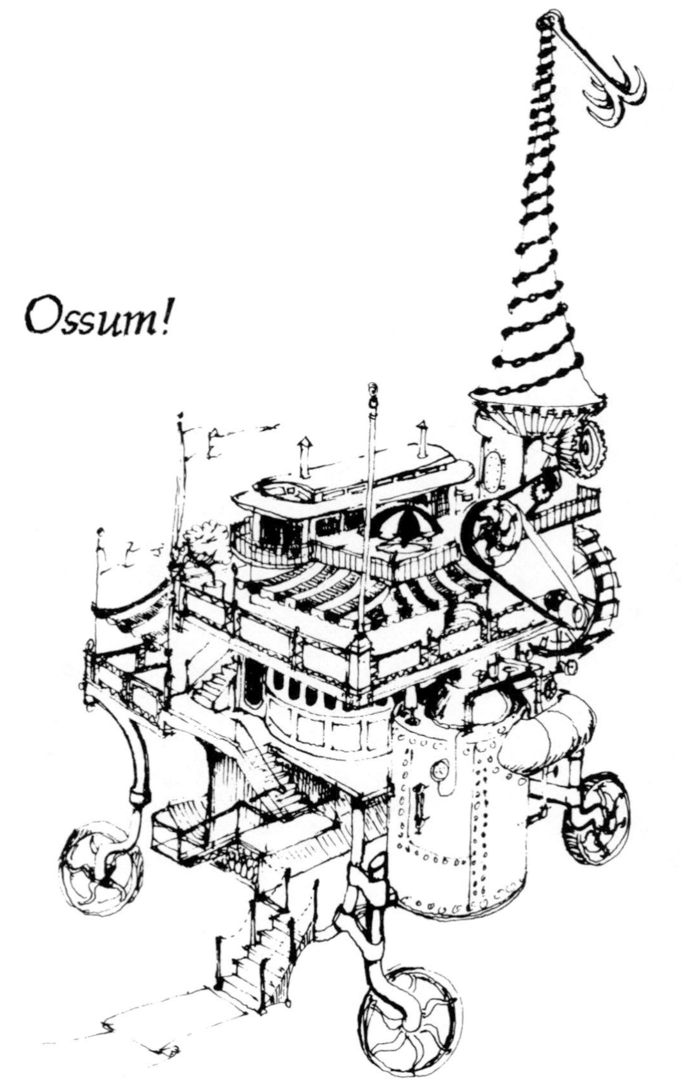

Today is the beginning of the Festival of Wood energy celebration. Regional winners of the National Energy Contest in the categories of tree production, fuel conservation, and combustion system design are being honored by officials representing the Department of Renewable Energy. The winners have received as their prize a luxury cruise aboard the amazing U.S.S. Ossum, a wood-fired terrestrial craft named after alternative energy pioneer Eugene Ossum.

The amazing Ossum stands before us. It is a four-story cylinder of cast iron and steel enclosing a boiler, a vertical mounted engine, a storage space for wood, and sumptuous accommodations for the passengers. An upward glance focuses on gilded capitals and clerestory windows through which the sun floats onto linen-covered tables. Each table carries a silver vase containing the perfect rose. Starched coated waiters glide silently about, catering to every whim and requirement of the solid-fuel heroes. Chilled champagne attends each table, and a canvas railed deck invites strollers. A leisurely game of quoits amid the talk of wood pellets and combustion temperatures helps to pass the time while waiting for the steam to build up.

Underneath this epitome of solid fuel technology are three swiveling castors that allow mobility in any direction.

Above the craft towers a massive, vertical cast shaft. Around it is wrapped 250 feet of the stoutest chain, and on the chain's end hangs a gigantic grappling hook.

The gang plank is withdrawn, the streamers and confetti are tossed, and as hats are thrown into the air and champagne corks pop, the towering mast begins to rotate under the power of the steam boiler. Visitors draw back to a safe distance as the mast spins faster and faster, becoming a blur. As the revolutions increase, the slender point of the shaft begins to wobble; the faster the turning, the more visible the strain. Suddenly it happens; the point of the shaft, burdened by the eccentric load of the hook, breaks off, throwing the chain in a graceful arch much as the fisherman releases line to a fish from his spinning reel. The grappling hook crashes to the earth, its tremendous weight and sharp points sink deeply, ripping up the blacktop with which we have covered the world. The chain hangs slack for a second following this release, but the spindle continues to rotate. The chain is still held by the unbroken lower portion of the mast, and the continuing whirl begins to recall the chain with a powerful winching action. The magnificent Ossum creeps slowly the 150 feet to its goal.

The brass band plays, and the crowd cheers the workers who halt the engine and scramble to the summit to replace the cast iron mast with a new one, refasten the chain, and begin the process once again.

Inside, the celebrated energy stars care little for the progress that has been made. The passenger appointments of this landlocked ship are so luxurious that there is no need to control the direction in which the hook flips; getting there isn't half the fun, it is all the fun. The passengers are content with this, and with the satisfaction that they are promoting a resource that the earth has always offered and will continue to offer.

These two creations, although amusing to contemplate, present an underlying question of a more serious nature. That is, to what extent is solid fuel an appropriate part of the technology of any given age? The flying steam carriage of the past demonstrates the pursuit of a goal that is unobtainable with the primitive technology of the day. The scenario of the future offers successful technology without an apparent purpose. Is one an improvement over the other? More relevantly, what does either example teach us as we sit by the protective warmth of our stoves?

The question is one of viability. Available technology can keep us in relative comfort with the turn of a dial. No woodlot management, cutting, stacking, splitting, or disposal of ashes is required. The use of wood or coal for primary, or even supplementary heating had passed from popular use as little as ten years before the creation of this book. Ten years hence, logic would tell us, stoves might again become quaint anachronisms hastening to a brief period in history when a nation panicked and thought it was running out of energy. Stoves will become as symbolic of one era as bomb shelters are of another.

The Green Mountains that surround us here in Vermont limit our line of sight in both literal and figurative ways. It is beyond our scope to discuss politics, economics or any of the other "issues" which affect the "big picture" insofar as the future of solid fuel heating is concerned. We do know stoves, however. We live with them, and frankly, we can't imagine life without them. If there is a future for stoves, then it relates to the fact that many others share our opinion that there is a viability to the idea of taking a local fuel and converting it to heat directly in the living space. Beyond any economic advantage this entails,

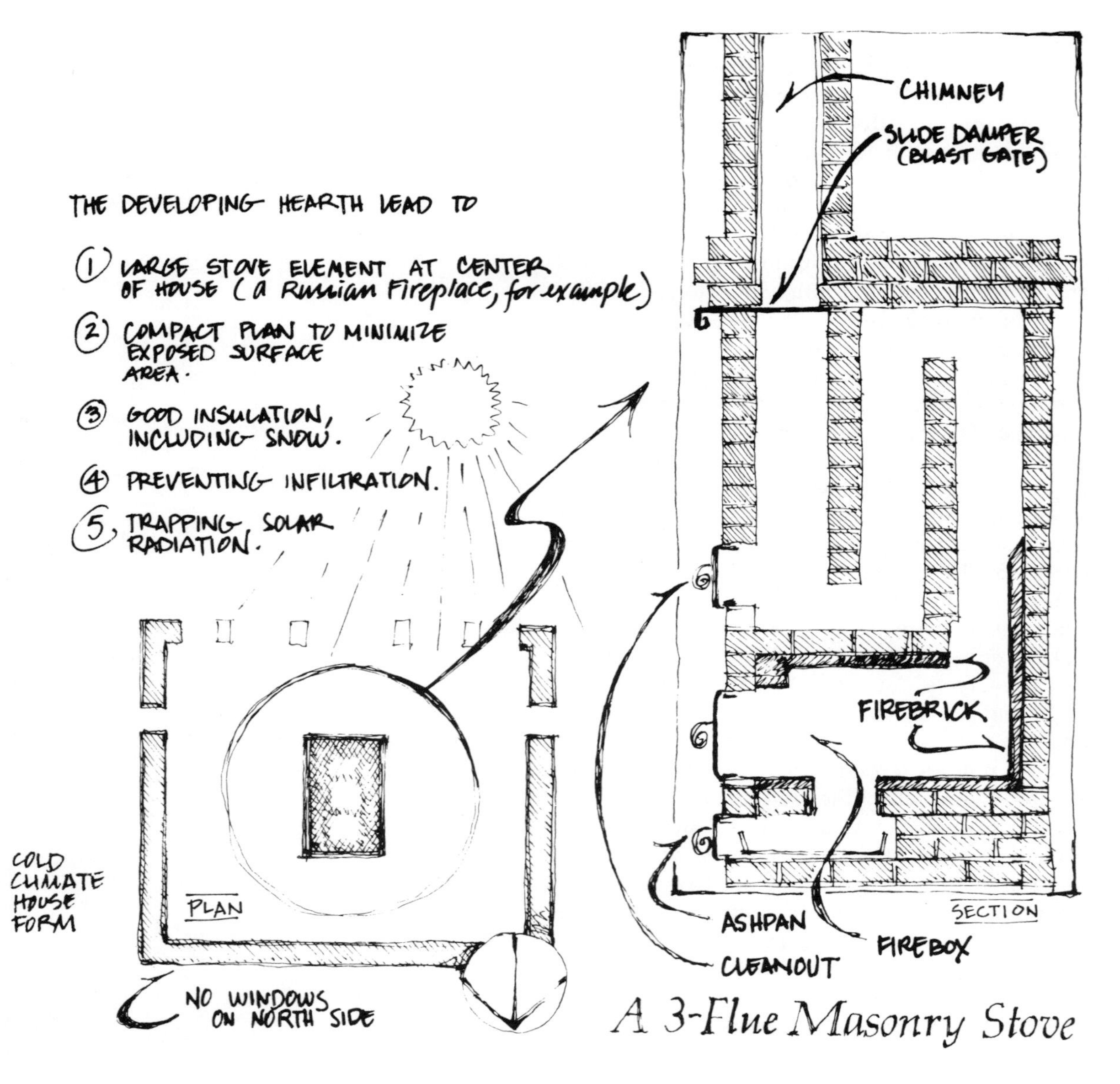

A 3-Flue Masonry Stove

there is an intangible benefit to having a focal point for warmth and security within the home. This benefit will continue to attract recruits to the ranks of stoveowners. A look at the past gives us the confidence to make this statement.

History of the Hearth

Imagine what it meant for Man to discover and to control the use of fire. He could then survive winters in extreme climates, thereby expanding his habitable range by quantum leaps. Moreover, he could establish a defendable sanctuary against both predators and the elements, forsaking forever nomadic ways. The fire served as a social catalyst as well. One can imagine among the earliest phrases of *homo sapiens* being the caveman equivalent of, "Hey, don't forget to stoke the stove before going to bed."

Communication that developed around the hearth led to the myriad accomplishments that are possible from the collective efforts of humans, from the domestication of grains to the construction of suspension bridges to landing a man on the moon. Surely, it is no coincidence that "hearth" is similar to "heart" in concept as well as in spelling.

Men tend to forget they are still naked apes, as unable to survive subzero temperatures now as a million years ago. The oil sheiks reminded us very graphically in the 1970's of our reliance on heat-producing technology. Without imported oil, many who took for granted the protection from the elements faced for the first time the prospect of going cold. The sudden knowledge of vulnerability caught many by surprise, as our "guarantee" of unlimited energy instantly became an illusion. In the manner of threatened animals we looked toward the sanctity of our hearths only to discover that they were no longer there or had fallen into disuse from years of neglect. Corrective action began immediately. Perhaps when historians look back on the 1970's one of the most positive world-wide social developments will have been the re-establishment of the hearth in the human home.

Technological Evolution

Between cavemen days and the late 18th century, the hearth made one major advance—it moved inside. One built a fire on the floor and poked a hole in the roof. The net result of smoke in the room and heat out the hole was inevitable, but still an improvement on the campfire. Man seemed to rest on his laurels at this point, as there were few improvements in the hearth for subsequent millennia. Even today the fireplace built in the contemporary home is likely to smoke and to dump most of its heat up the chimney.

Two Americans, both named Benjamin, used Yankee ingenuity in the late 1700's to come up with dramatic advances in the cause of home heating. Each looked at the perennial problems of fireplaces—smoking and heat loss—and independently came up with solutions.

Benjamin Franklin in 1740 enclosed the fireplace in cast iron, thereby inventing the freestanding stove with closing doors that to this day bears his name.

Benjamin Thompson, better known as Count Rumford, refined the principles of fireplace construction to levels of sophistication as yet un-

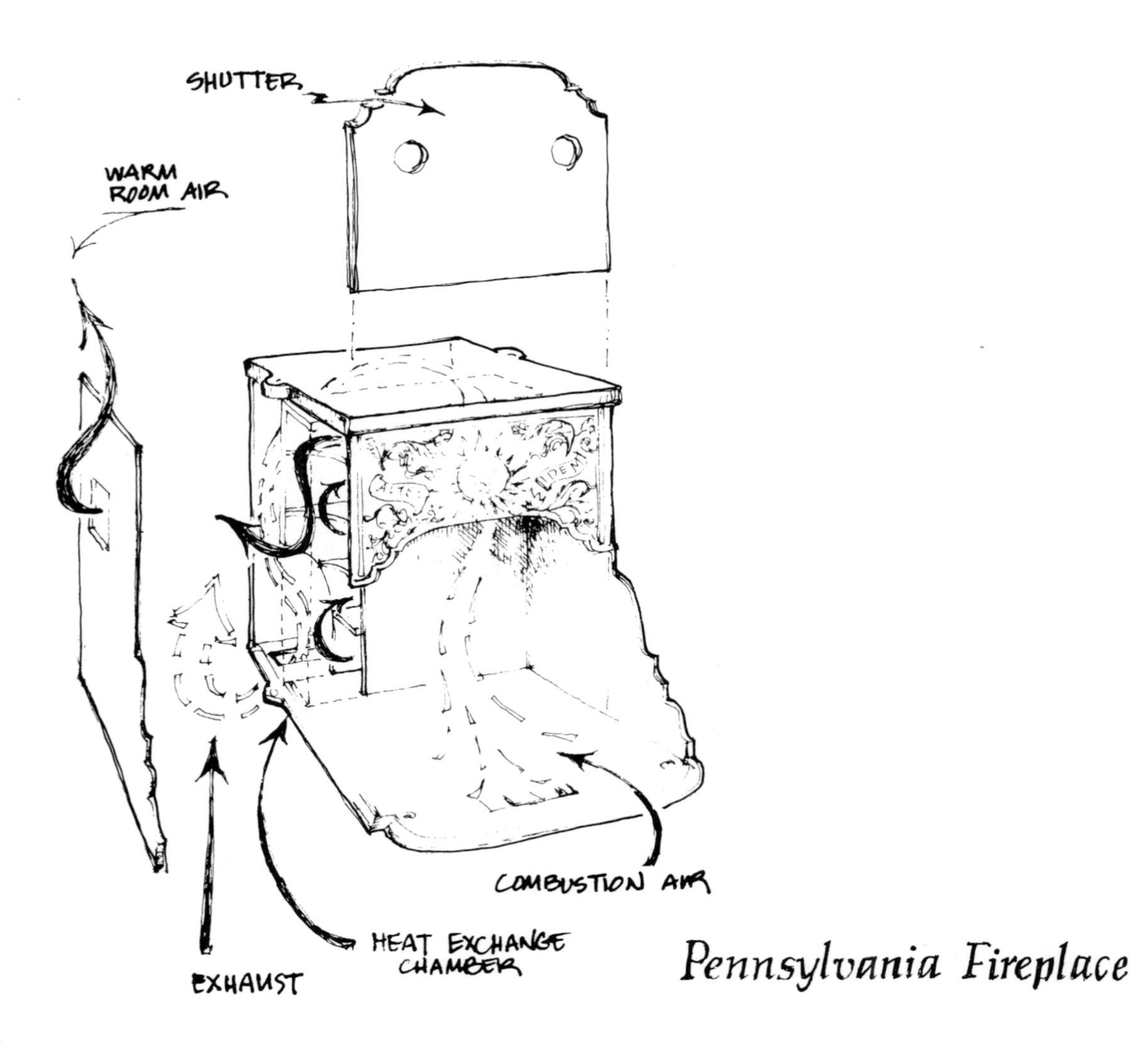

Pennsylvania Fireplace

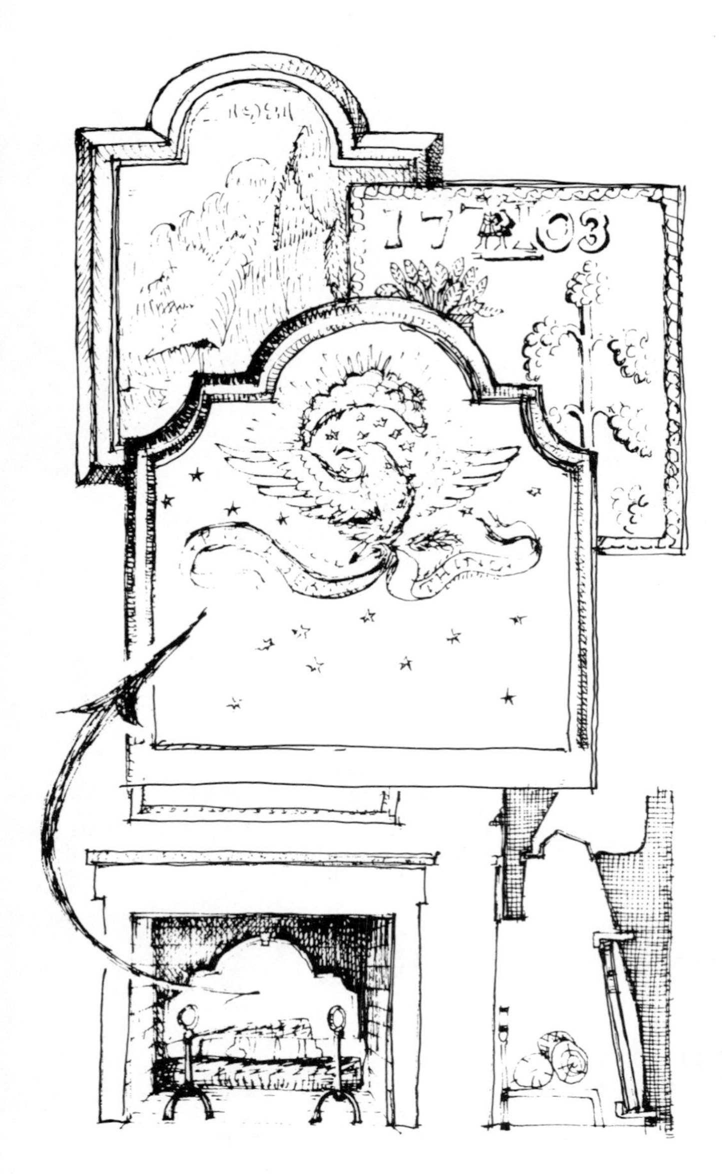

FIREBACKS CAME INTO USE, AT LEAST 500 YEARS AGO, TO PROTECT THE BRICKS AND MORTAR OF THE FIREPLACE FROM DETERIORATION. WITH MODERN HIGH TEMPERATURE MORTARS, THEIR MAIN FUNCTION TODAY IS TO RE-RADIATE THE FIREPLACE HEAT.

The Two Benjamins

The two Benjamins were contemporaries in Colonial America during an era when a clever man could be the master of many trades. They were native Yankees whose primary achievements occurred only after each had left his native Massachusetts. One was apprenticed to a printer, the other to a physician, but both had interests too diverse to be bound by a single discipline. The one subject to which they individually addressed their individual talents was the inefficiency of the common fireplace.

Benjamin Franklin's approach to the fireplace was to enclose the fire in a freestanding unit made of a material such as cast iron which would transfer radiant heat. Limited air flow would mean controllable combustion and therefore greater efficiency. A testimony to the validity of Franklin's innovation is readily apparent from seeing how many contemporary stoves pay obvious homage to the styling of the stove which bears his name.

Benjamin Thompson, on the other hand, refined the design of fireplaces according to the simplest principles of geometry. In the nearly two hundred years since his death there has arguably been no improvement in the art of fireplace design. In fact, many fireplaces made in the last one hundred years have been constructed with an eye toward aesthetics, often at the expense of function. Benjamin Thompson, whose title as Count Rumford is imprinted as indelibly on a type of fireplace as Franklin's name is on his stove, would turn in his grave to see the fieldstone, raised-hearth monstrosities which violate every tenet of his classic design.

Each man boasted accomplishments which far outshone his contribution to the hearth, yet while Franklin lives on as the embodiment of the inventive American spirit, Thompson's accomplishments languish in obscurity. The reason is simple: When the American Revolution came, Franklin chose the side of the rebels, eventually adding the credits of patriot and statesman to his reputation as an inventor.

Thompson, despite an equally distinguished career spanning diplomacy and the humanities, always retained his loyalties to George III, thereby being branded as a Tory and forever relegated to footnote status in American history. The principles of combustion and air movement, happily, are blind to politics, and Count Rumford's fireplaces still remain as the highest expression of their art.

surpassed. Franklin, then, solved the smoking fireplace problem by boxing it while Count Rumford outsmarted it. Although an argument could be made for Rumford's solution being the more aesthetic, Franklin's was the more economical. By closing the front doors of the fireplace one controlled the combustion, resulting in less wood consumption and longer burn times. Thanks to the efforts of the two Benjamins, both the function and spirit of the hearth entered new eras.

Franklin saw the importance of his innovation and in a gesture of magnanimity inconceivable by today's standards, voluntarily refused to patent it, stating:

> That, as we enjoy great advantages from the inventions of others, we should be glad of an opportunity to serve others by any invention of ours; and this we should do freely and generously.*

Despite his beneficence, Franklin's "Pennsylvania Fireplace" did not come into widespread use until there was a compelling economic reason for people to change more wasteful ways. This came in the early 1800's. With an unquenchable zeal to tame the land for agricultural purposes, colonists cleared out vast sections of the Northeast in one of the great collective forestry mismanagements of all time. The resulting wood shortage motivated the population, especially those living in more urban areas, to investigate more efficient ways of burning wood as well as

* Jared Sparks, ed., *Works of Benjamin Franklin* (Boston, 1838), in *Cast With Style*, Tammis Kane Groft (Albany Institute of Art, 1981), p. 13.

① SQUARE FIREBACK IS ABOUT 15" WIDE.

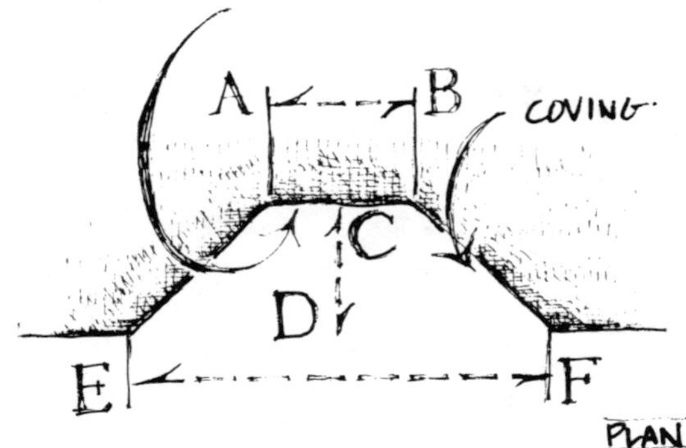

② FIREPLACE DEPTH CD IS EQUAL TO FIREBACK WIDTH AB.

③ FIREPLACE OPENING WIDTH EF IS THREE TIMES FIREBACK WIDTH AB.

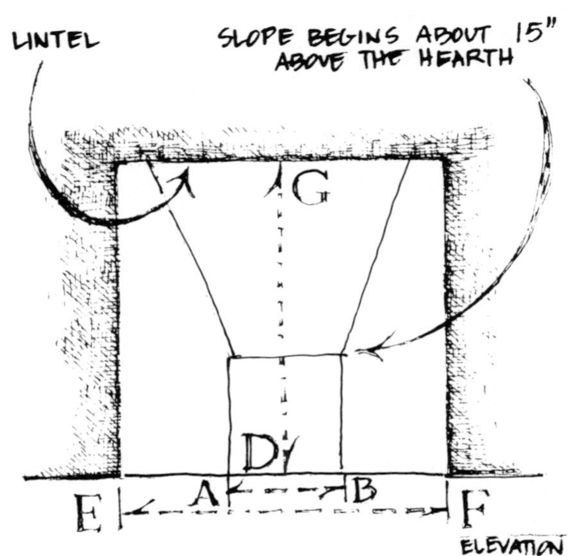

④ FIREPLACE OPENING HEIGHT DG IS THREE TIMES FIREBACK WIDTH AB AND EQUAL TO FIREPLACE OPENING WIDTH EF.

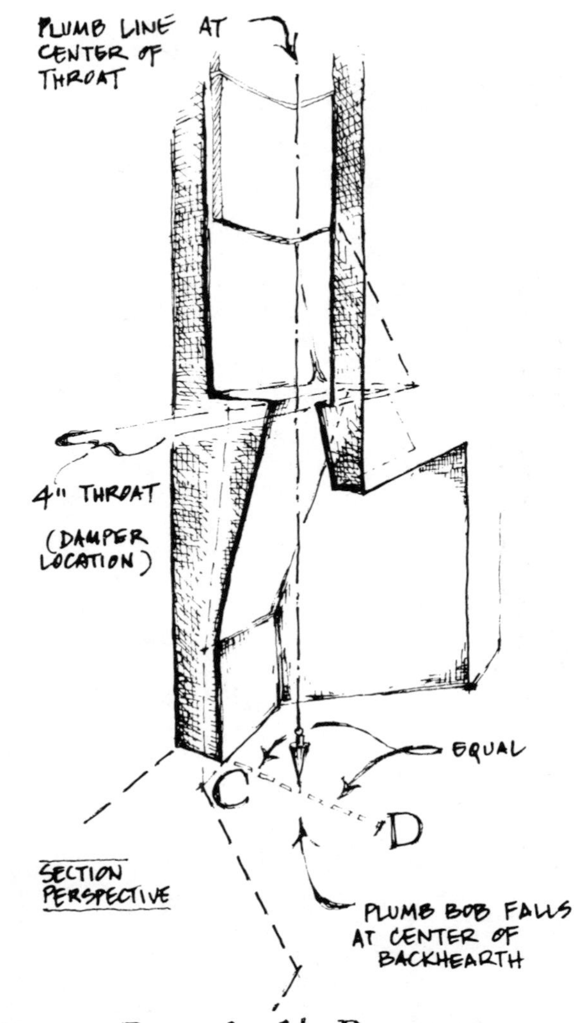

Count Rumford's Proportions

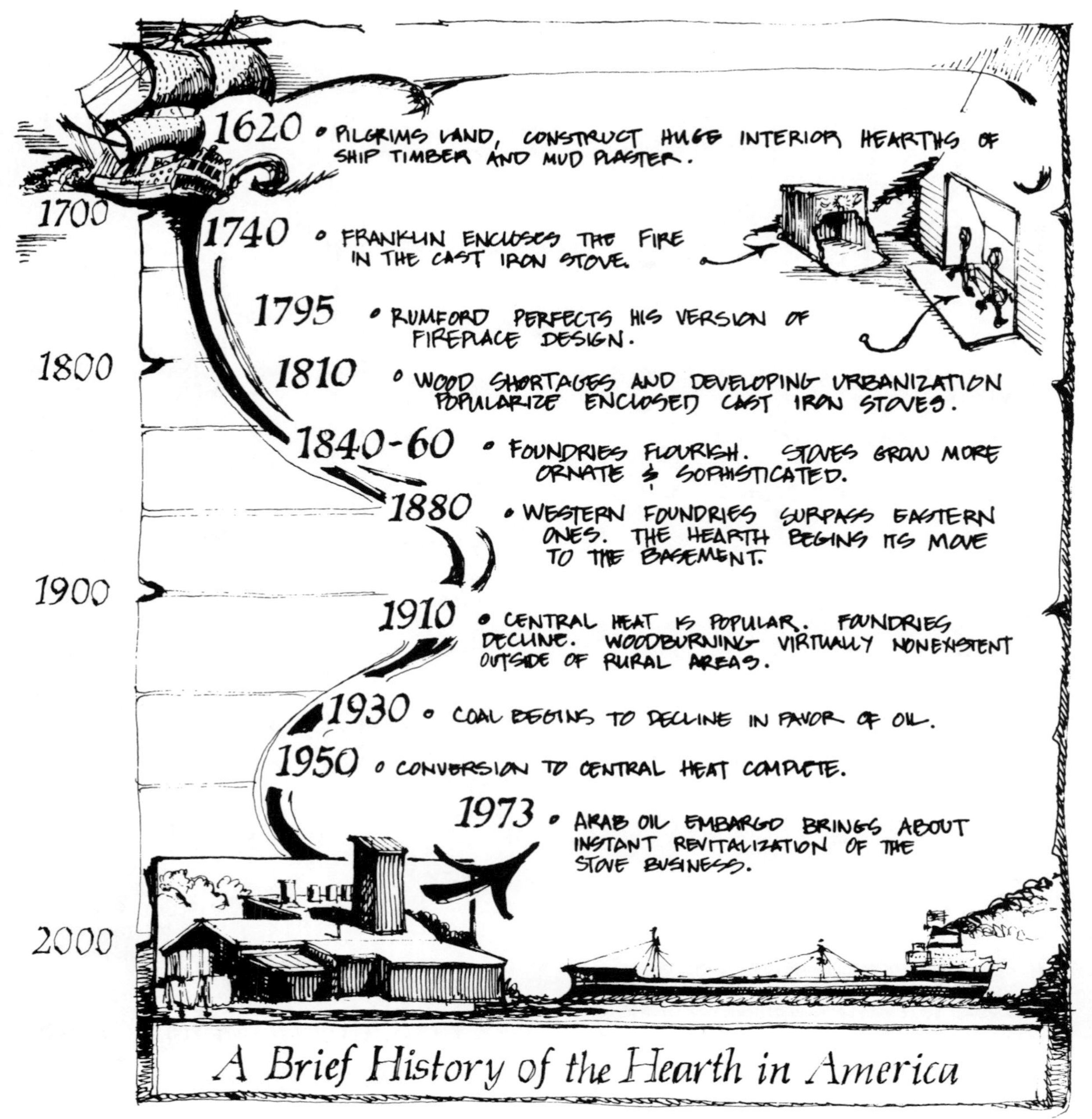

A Brief History of the Hearth in America

the use of alternative fuels. Before long, the wood/coal, freestanding cast iron parlor stove became a standard in many American homes.

It is inevitable that innovations in the hearth will have far reaching effects on other aspects of life. House design was soon radically altered as if to take advantage of the new way of conserving resources. Ironically, dwellings became larger, more airy, and less heat-efficient. High ceilings and full second stories allowed better circulation and use of a stove's heat, yet often at the expense of fuel consumption. The lesson here is that the squandering of resources is inevitable. Technological advances only provide new and creative ways to be wasteful.

Parallels in contemporary history abound. Public reaction to the Arab oil embargo of 1973 attached a new importance on the development of alternative energy sources. Solar, wind, and solid fuel technologies suddenly became attractive possibilities. Fortunately, the westward movement of the agricultural business combined with several generations of burning fossil fuels had resulted in a reforestation of much of the Northeast. Through benevolent neglect wood was now abundant, and a major resource was discovered. Once the impact of the initial oil shortage had passed, however, Americans quickly reverted to consumption habits developed when oil resources were thought to be infinite. Big cars again became fashionable, and techniques of fuel conservations were quickly forgotten. Another shortage in 1979 again showed the fragility of the fossil fuel supply and brought another minor boom in the development of alternative energy sources. Such periods of boom and bust, glut and shortage can be expected as long as there are fluctuations in the fuel market. For what it is worth, market analysts are united in the view that for the forseeable future the fuel supply will continue to be unstable.

When Vermont Castings opened its computerized foundry in 1979, a member of the staff undertook research to discover when the last foundry constructed for the pouring of stove plate had been built in this country. No records of any such project could be found as far back as the turn of the century. More surprising than the revival of the stove industry in the 1970's, however, is a perception of the depths to which it had declined since its glory days.

Modern Times

As an inventive *tour de force* Franklin's stove was overshadowed by Rumford's fireplace innovations, but external factors eventually tipped the scales heavily in favor of the more fuel-efficient cast iron stove. Throughout the first half of the 19th century the foundry industry flourished. As many as thirty-two separate, independent foundries, for example, operated simultaneously in the Troy–Albany section of New York. Stoves were available in a variety of designs and styles that would boggle the mind of today's prospective stove buyer. Designers and craftsmen competed to come up with increasingly elaborate and intricate designs to capture the fancy of an ever-whimsical and more discriminating consumer. Stoves became more than heaters. They were pieces of furniture, they were centerpieces; they were works of art.

The Sargent, Osgood & Roundy Co.

When Vermont Castings was formed in 1975 the founders chose the site of the defunct Sargent, Osgood & Roundy Co. foundry for two reasons, One, it was fitting that a fledgling stove company be located on the premises where fifty years earlier stoves had been cast, and two, it was the cheapest space available, just right for a company founded on little more than a good idea and a sense of style.

It has always felt right somehow to be located on the old foundry site. We do not know many details of the original business, other than that their primary products were farm implements rather than stoves. Occasionally an old timer stops by to swap a few yarns, or an artifact comes up at a local yard sale or auction, and we can fill in a piece of the puzzle. But mainly the business lives on every morning when we come down the curved driveway which is Prince Street, and we see the red brick, rectangular smoke-stack which bears the faded, white letters The Sargent Roundy Corporation. Squint your eyes, and it could be 1890.

The Troy-Albany Stovemaker's Art

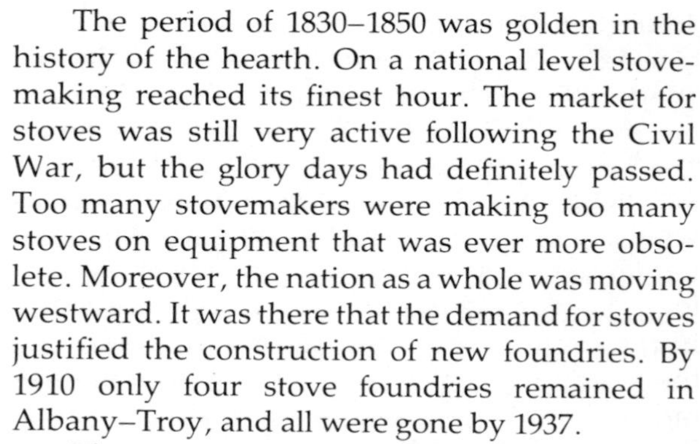

The period of 1830–1850 was golden in the history of the hearth. On a national level stovemaking reached its finest hour. The market for stoves was still very active following the Civil War, but the glory days had definitely passed. Too many stovemakers were making too many stoves on equipment that was ever more obsolete. Moreover, the nation as a whole was moving westward. It was there that the demand for stoves justified the construction of new foundries. By 1910 only four stove foundries remained in Albany–Troy, and all were gone by 1937.

There were more contributing factors to the decline of the stove industry. New fuels derived from petroleum made inroads into the market. Decidedly unromantic, these fuels were relegated to the basement to serve a strictly functional purpose. As happened with the advent of the Franklin stove, central heating enabled new and more whimsical turns in home design. The fanciful flights of Victorian imagination were given birth by the practical reality that central heat could keep a larger space warm.

Fireplace design, it should be mentioned, declined steadily. To this day there are few masons who understand the efficiency implicit in Thompson's simple principles. By 1900 the fireplace had become purely an aesthetic object, serving the social purposes as a gathering point, but divorced from its functional mandates. Even the picturesque fireplace of the contemporary ski lodge created during an era of heightened energy consciousness offers no improvement over the Rumford design. Its raised hearth, fieldstone surface, glass doors, and blower heat exchangers are merely bandaids to partially compensate for a basic abomination of design. Examine the lintel and you will likely see the stains of escaping smoke, the very problem Thompson set out to correct nearly two centuries ago.

With the relegation of the parlor stove to the basement in the form of the furnace or boiler, what became of the social function of the hearth? Largely, the void was filled by a series of technological surrogates, a development that continues in parallel with the contemporary renaissance of the hearth. First, the family gathered around the radio, then the televison which bathed us in its flickering, bewitching blue light. Now, many homes are not complete without a "home entertainment center," a space-age masterpiece in which the proverbial "boobtube" is enhanced by the videotape recorder, laserdisc, home computer, and Space Invaders console. It is not surprising that the companies who make these products depict the family securely plugged into their electronic hearth attached by the umbilical of the hand-held control unit. The caveman wandering into the 20th century home one chilly evening would recognize a familiar scene and back up to the lit screen to catch some expected warmth.

Other signs of hearth substitution abound. Man the Hunter comes home from a hard day at the office, dons his ritual uniform, and marches out to the backyard gas grill from which he dispenses chunks of charred meat. In our local taverns and pubs (a term derived from "public house") we can expect the latest in hearths. We find it too frequently in the form of the mesmerizing television.

Whether or not Pac-Man provides the same security and stimulation as a bed of glowing embers could be the subject of spirited debate. The proponents of simplicity could rightly point to the fact that the supplanting of the hearth by electronic gadgets has coincided with the deterioration of the family unit in America. For evidence of the timeless and universal appeal of the fire, one can flip through any glossy magazine to see how frequently Madison Avenue uses flickering flames to create a comfortable ambiance as a backdrop for selling products ranging from cognac to stem crystal. And when a President wants to come into our homes to deliver a heartfelt message, it is no coincidence that the crackle of a fireplace comes with him.

The Electronic Hearth

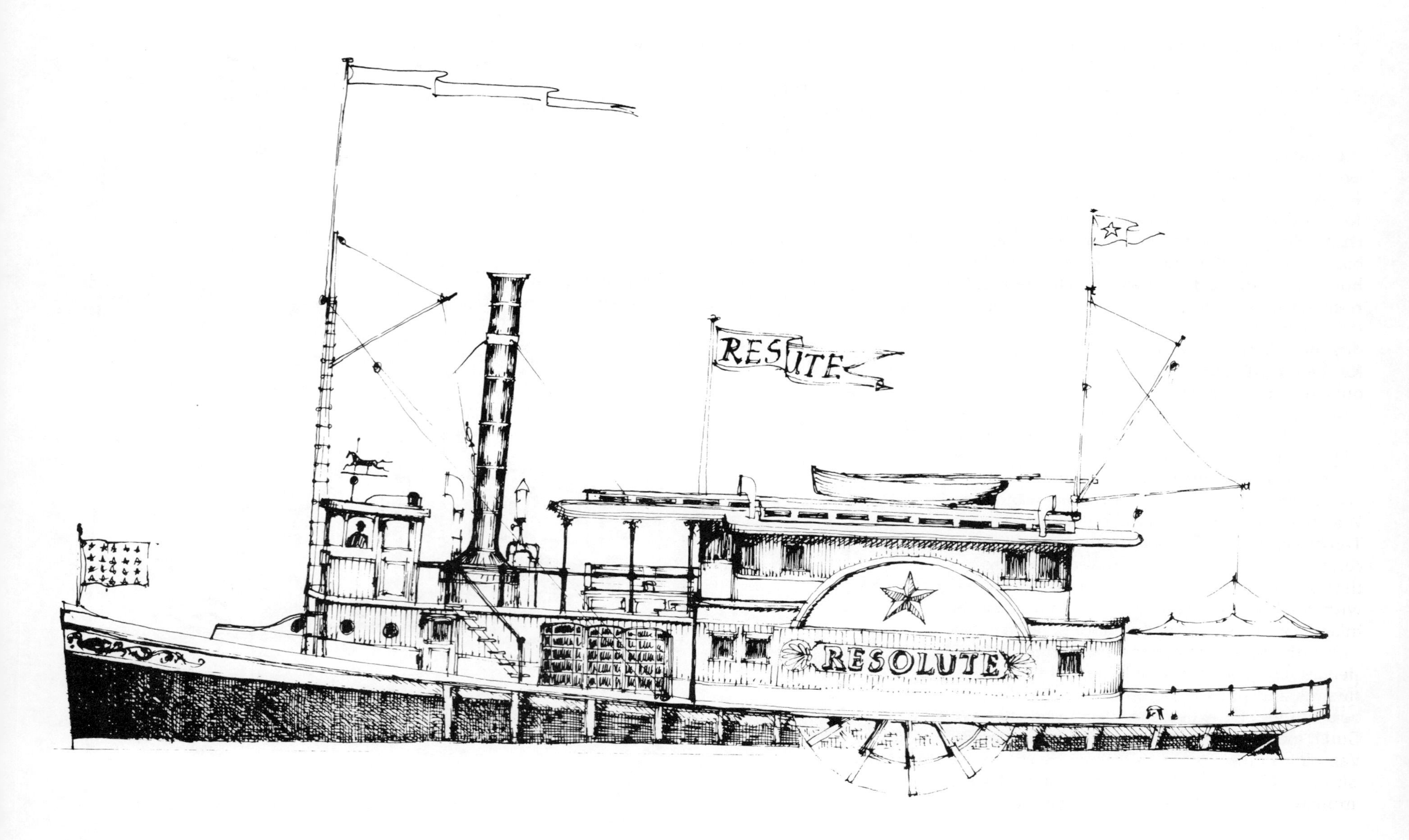
RES UTE
RESOLUTE

The ideal hearth is equal parts function, beauty, and magic. The sociological implication of its rise, fall, and resurrection in America is a perfect topic for discussion when the family is together after Thanksgiving dinner. Kindle up the fireplace or stove and watch the flames lick upwards in an infinitely varied display. Words will either come naturally or seem unnecessary.

The future undoubtedly holds more for us in the way of home computers, video games, and other components of the electronic hearth. Let's hope that this does not exclude the continued reinstitution of the real hearth as a focal point in the American home, for—who knows?—someday the lights may go out, and we will be grateful for the warmth and glow. After all, a home without a hearth is really not a home at all.

The S.S. Resolute

We are not sure that this ship will ever be built. Tentatively dubbed the *S.S. Resolute*, it is conceived as a wood/coal steam-fired paddle wheeler that would enable us to visit our customers within calling distance of the coast or a major inland waterway.

We picture ourselves sitting on the fantail steaming westward into Long Island Sound, steering a course of 260° magnetic. Off the port bow is Orient Point, Plum Island, and the Two Gulls. New London is to the starboard, and the Valiant Rock Bell and Block Island are fast diminishing astern. Steam billows from the stack and immediately dissipates to reveal the blueness of the Indian Summer sky. There is a suggestion of fall in the air, but in the strong sunlight summer reigns supreme. The brass is polished, the deck is swabbed . . . even the soaring gulls are polite.

The *S.S. Resolute* has not yet been built for a million reasons, primary among them being the knowledge that to make a moment like this possible, an infinity of barnacle scraping, deck varnishing, and engine tinkering are required. So it is with stoves. The idle, secure moment of reverie from whence a great thought is born is made possible only by the investment of many hours of tedium. The rituals of cutting, hauling, and stacking comprise a litany which somehow becomes greater than its components, and makes it all worthwhile.

But hope springs eternal. Someday the rigors of stove building and preparing for winter will subside long enough to let us undertake the *S.S. Resolute*. Until then, whenever you are by the shore, watch for the telltale steam and listen for the whistle.

Winter Checklist

The stove has had a few shakedown cruises in early fall. Then there is the cold evening when you burn it overnight to yourself the trouble of re-kindling. The next thing you know the stove has been operating continuously for five months. To shut it down for any reason constitutes a hardship, and it will require constant care and tending. Pay attention, for the line between devotion and bondage is a fine one indeed.

□ ***Check your insurance policy.*** *No kidding. Some insurance companies have passed prohibitive rules regarding stoves in the home. If this is true of yours, don't change the stove, change insurance companies. Most insurance companies, however, provide sound safety information, so pay heed.*

□ ***Learn the good points of cabin fever.*** *You don't have any choice anyway.*

□ ***Order seeds.*** *The seed catalogues comes out just after the New Year. Spend lots of time hanging around the stove planning your springtime activities.*

□ ***Sharpen you chainsaw.*** *The winter is a great time for puttering. You will have enough frustration with your chainsaw in the spring. You might as will begin the season with your unit in tip-top shape.*

□ ***Bag your ashes.*** *The safest method we've heard of for disposing of wood ash is transferral from stove to galvanized can to plastic bag. Allow two full weeks in the can for ashes to completely extinguish. Coal ashes go from stove to metal can to land fill. No one has yet come up with a practical use for them.*

□ ***Don't burn Christmas wrappings.*** *Or the dried Christmas tree. These are sure chimney fire starters.*

□ ***Check fuel supply on Groundhog Day.*** *Oldtimers say you should have half your fuel supply left as of this day—wood or coal. See how well you estimated your needs.*

□ ***Develop a ritual.*** *Once your stove is operating there is a tendency to "let sleeping dogs lie." This can be dangerous unless you force yourself to check your pipes, hearth, and chimney on a regular basis. Hopefully the checks will reveal nothing alarming, but the consequences are great enough to justify the effort.*

□ ***New Years' resolutions.*** *By now you have hauled enough wood or coal to realize how important the smallest shortcut is in your logistic set-up. You've learned the difference between waking up to a cold stove and hot coals. You may even know the pleasures of climbing onto a snowy roof at twenty below to clean a chimney. These lessons are only as valuable as the changes they effect in your planning for the next heating season.*

□ ***Philosophize.*** *You are living a lifestyle that requires an effort which is infinitely more work than that which is technologically possible. There are economic justifications, to be sure, but basically you bave chosen to live a certain way. Sit by the stove, and think it through.*

□ ***Adjust expectations.*** *The same stove which produced a 10 hour burn in November will now consume a load of fuel in 6 hours. The reason is simple—more BTU's are needed to keep your house comfortable.*

□ ***Watch the news.*** *The world continues in turmoil. Take some assurance in the fact that as long as you have a stove and fuel, you will be warm.*

Appendix

Stove Shops

There are many quality stove shops in the United States and Canada. These are some with whom we have had personal experience, and we can vouch for their orientation toward safe and responsible stove operation.

ALABAMA

Summerwood Stove Company
2345 Whitesburg Drive
P.O. Box 60
Huntsville, AL 35801

ALASKA

Northeat Woodstove
1306 Chugach Way
Anchorage, AK 99503

The Woodway
918 College Street
Fairbanks, AK 99701

The Firebox
883 Basin Road
Juneau, AK 99801

ARIZONA

The Energy Center
2004 East Santa Fe
Flagstaff, AZ 86001

The Fire House
520 West Sheldon Street
Prescott, AZ 86301

ARKANSAS

Woodwise Enterprises
4304 Pike Avenue
North Little Rock, AR 72116

CALIFORNIA

Gardenway-Sacramento
7654 Green Back Lane
Citrus Heights, CA 95611

Sierra Timberline
525 East Main Street
Grass Valley, CA 95945

Energy Unlimited
147 West Richmond Avenue
Richmond, CA 94801

Energy Unlimited
River Street
Truckee, CA 95734

Real Goods Trading Company
308 East Perkins Street
Ukiah, CA 95482

Real Goods Trading Company
358 South Main Street
Willits, CA 95490

Firehouse West Inc.
8059 Aptos Street
Aptos, CA 95003

Firehouse West
790 South Winchester Blvd
San Jose, CA 95128

Forden's
857 Monterey
San Luis Obispo, CA 93401

Stove Works Energy Center
1440 Hartnell Avenue
Redding, CA 96002

Julian Lumber
2125 Main Street
Julian, CA 92036

San Bernadino Fireplace
Specialties
457 West Highland Avenue
San Bernadino, CA 92405

Okell's Fireplace
134 Pacific Coast Highway
Hermosa Beach, CA 90254

COLORADO

Mountain Stove Works
942 Pearl Street
Boulder, CO 80302

Mountain Stove Works
505-30 Road
Grand Junction, CO 81501

CONNECTICUT

Pete's Cycles and Stoves
934D Post road
Guilford, CT 06437

Pete's Cycles and Stoves
51 College Street
New Haven, CT 06510

Pete's Cycles and Stoves
873 Post Road
Old Saybrook, CT 06475

FLORIDA

The Wood Stove Store
457½ West Virginia Avenue
Tallahassee, FLA 32301

The Wood Stove
2222 Northwest 6th Street
Gainesville, FLA 32601

GEORGIA

Gardenway Living Center
2744 South Cobb Industrial Boulevard
Smyrna, GA 30080

HAWAII

The Most Irresistible Shop in Hilo
110 Keawe Street
Hilo, HI 96720

IDAHO

Woodstove Works
Main Street
Box 249
Oakley, Idaho 83346

ILLINOIS

Resource Alternatives
P.O. Box 175
Old Galena Road
Mossville, IL 61552

Millhouse Center
402 North Race
Urbana, IL 61801

INDIANA

Battleground Stovepipe
Railroad Street
Battleground, IN 47920

The Practical Answer
525 Lincoln Way West
South Bend, IN 46601

Battleground Stovepipe
4201 West 62nd Street
Indianapolis, IN 46268

Energy Alternatives
County Road 300 North
Mongo, IN 46771

IOWA

Ralston Creek Stove and Tool, Inc.
320 East Benton
Iowa City, IA 52240

Ralston Creek Stove and Tool, Inc.
10 Main Street
Bondurant, IA 50035

KANSAS

The Firebox
4900 10th Street
Great Bend, KS 67530

Woodstoves Inc.
615 Massachusetts Street
Lawrence, KS 66044

KENTUCKY

Eastwood Stove Company
16201 Eastwood Cutoff
Eastwood, KY 40018

MAINE

Vermont Castings, Inc.
53 Kennebec Street
Portland, ME 01401

MARYLAND

The Energy Store
9130 Red Branch Road
Columbia, MD 21045

Patamoke Stove Works
Route 1
Box 59
Trappe, MD 21673

MICHIGAN

Great Lakes Energy Systems
109 Water Street
Boyne City, MI 49712

Great Lakes Energy Systems
5690 US 31 North
Acme, MI 49610

Heat and Sweep
706 South Main Street
Plymouth, MI 48170

Heat and Sweep
119 South Putnam Street
Williamston, MI 48895

MINNESOTA

Energy Updaters
109 Washington
Brainerd, MN 55401

Ironwood
115 North First Street
Minneapolis, MN 55401

Energy Plus
3000 Miller Trunk Highway
Duluth, MN 55811

MISSISSIPPI

Ben's Woodheat
P.O. Box 571
610 Howe Street
McComb, MS 39648

MISSOURI

Rockwood Stoves
158 West Argonne
Kirkwood, MO 63122

MONTANA

North Country Stove Works
40 South Park
Helena, MT 59601

The Fuelmizer
405 West Idaho
Kalispell, MT 59901

NEW HAMPSHIRE

Sandhill, Inc.
36 Grove Street
Peterborough, NH 03458

NEW JERSEY

Chimney Sweep Energy Barn
Route 23 South
Newfoundland, NJ 07435

Scandamerican Energy
Systems
U.S. Route 9
Staffordville, NJ 08092

NEW MEXICO

The Firebird
330 Montezuma
P.O. Box 5275
Santa Fe, NM 87502

NEW YORK

Chimney Sweep
RD #3, Box 175A Shaw Road
Middleton, NY 10940

Southampton Stove Company
West Main Street
Southampton, NY 11968

Williamson Hardware
Four Corners
Williamson, NY 14589

Fingerlakes Fabricating
726 West Court Street
Ithaca, NY 14850

NORTH CAROLINA

Solar P.I.E.
P.O. Box 506
Highway 108
Columbus, NC 28722

Triad Energy Systems
Bull Durham Energy Systems
2000 Chapel Hill Road
Durham, NC 27707

Triad Energy Systems
Sunspot Energy Systems
118 East Main Street
Carrboro, NC 27510

Stoves 'N' Such
P.O. Box 12897
4240 Kernersville
Winston-Salem, NC 27107

OHIO

Bauer Stoves
3548 State Route 54
Urbana, OH 43078

The Alternative
1028 Bridge Street
Ashtabula, OH 44004

Western Reserve Stove Store
12385 Kinsman Road
Newbury, OH 44065

OKLAHOMA

The Energy Store
203 South Porter
Norman, OK 73070

OREGON

Larson-Thomas and Company
411 High Street
Eugene, OR 97401

Gardenway Living Center
1802 Jantzen Beach Center
Portland, OR 97217

Cascade Heat With Wood
1559 Dowell Road
Grants Pass, OR 97526

PENNSYLVANIA

Stove Shop
Route 29 and Pothouse Road
Phoenixville, PA 19460

Landscape II
201 Elmwood Street
State College, PA 16801

Stoves 'N' Stuff
Route 309
Tamaqua, PA 18252

John Wright Foundry Store
Box C-40 North Front Street
Wrightsville, PA 17368

The Fireplace
1651 McFarland Road
Pittsburgh, PA 15216

RHODE ISLAND

Economy Ornamental Works
464 Maple Avenue
Barrington, RI 02806

TENNESSEE

The Fireside Shop
3909 Martin Mill Pike
Knoxville, TN 37920

Fuel Miser Wood Heater
Company
4924 Highway 61 South
Memphis, TN 38109

VERMONT

Gardenway Living Center
186 Williston Road
Burlington, VT 05401

VIRGINIA

Nova Wood Heat
10120 Colvin Run Road
Great Falls, VA 22066

The Taproot
14501 Warwick Boulevard
Newport News, VA 23606

The Taproot
515 York Street
Williamsburg, VA 23185

WASHINGTON

Sutter Home Woodstove
Company
5333 Ballard Avenue NW
Seattle, WA 98170

The Stove Shoppe
East 1524 Sprague Avenue
Spokane, WA 99202

Ashes Quality Stove Shop
512 South 3rd Street
Yakima, WA 98901

WISCONSIN

Eder Nursery
5300 Highway K
Franksville, WI 53126

Energy Unlimited
1821 East Mason
Green Bay, WI 54302

Eder Nursery
4917 West Fond du Lac
Milwaukee, WI 53216

Applewood Stove Works
Route 3
Polynette, WI 53955

Eder Nursery
N 8 W 24040 Bluemound
Waukesha, WI 53186

The Cozy Hearth
213 North Main
Rice Lake, WI 54868

Woodshed Stove Works
1402 West Court Street
Janesville, WI 53545

CANADA

ALBERTA

The Fireside Store Ltd.
1026 124th Street
Edmonton, ALTA T5N1P6

BRITISH COLUMBIA

Woodburn Stoves and Fuel Ltd.
110 Fell Avenue
North Vancouver, BC V7P2K1

Fireplace Center Ltd.
1200 Battle Street
Kamloops, BC V2C2N6

Victoria Fireplace Shop Ltd.
1408 Blanshard Street
Victoria, BC V8W2J2

Duncan Hearth and Home Ltd.
531 Canada Avenue
Duncan, BC

Country Hearth and Home Ltd.
#14 Pinetree Square
1708 Bowen Road
Nanaimo, BC V9S1G9

MANITOBA

Added Energy
2555 Pembina Highway
Winnepeg, MAN R3T2H5

NEW BRUNSWICK

Alternate Heating Ltd.
621 Rothesay Avenue
St. John East, NB E2H2G9

NEWFOUNDLAND

Home Heat Services—
Ultramar Canada Ltd.
182 O'Leary Avenue
St. John's, NFLD A1C5T5

Ultramar Canada Ltd.
Main Street
Windsor, NFLD

Ultramar Canada Ltd.
Transcanada Highway
CornerbrooK, NFLD

NOVA SCOTIA

Energy Alternatives
2 Croft Street
Amherst, NS B4H4B8

Home Energy Audit, Ltd.
653 George Street
Sydney, NS B1P1L2

The Woodstove Store Ltd.
5217 Blowers Street
Halifax, NS B3J1J5

Central Supplies Ltd.
P.O. Box 1382
Transcanada Highway
Antigonish, NS B2G2G7

ONTARIO

Alternative Enterprises
RR 4, 1 Highway 28
Peterborough, ONT K9J6X5

The Avenue Road Woodstove
Store
174 Avenue Road
Toronto, ONT M5R2J1

The Efficient Woodstoves
192 A Main Street West
Huntsville, ONT P0A1K0

Cameron's Insulation Ltd.
Eamer's Corners
Cornwall, ONT K6H5R6

House By The School
98 Victoria Avenue
Vineland, ONT

Rainbow Woodstoves
139 Dunlop Street East
Barrie, ONT L4M1B1

Rainbow Woodstoves
85 Broadway
Orangeville, ONT L9W1K1

Rainbow Woodstoves
Junction of Highways 10, 24, 89
Shelburne, ONT L0N1S0

The Source
423 Colburne Street
Brantford, ONT N3S3N6

Tidman's Furniture and Appliance
166 Leslie Street
Newmarket, ONT L3Y3E4

Wood 'N' Energy
McDonald's Corners
ONT K0G1M0

Wood 'N' Energy
214 Dalhousie Street
Ottawa, ONT K1N7C8

Wood Power
123 Woolrich Street
Guelph, ONT N1H3V1

The Woodstove
25 Center Street
Strathroy, ONT N7G1T5

QUEBEC

Malvina Enterprises
6 Queen Street
Lennoxville, QUE J1M1Z3

Millete Woodstoves
280 Grand Cote
Rosemere (Montreal)
QUE J7A1J6

Treg Le Ramoneur
540 Rue Bagot
Quebec City, QUE G1N2A4

La Boutique Du Foyer Inc.
De La Mauricie
5375 des Forges
Tres Rivieres, QUE G8Y5L5

Foyer Universal Inc.
2361 Boul Rosemont
Montreal, QUE H2G1T9

Foyer Universal Inc.
6185 Boul Taschereau
Brossard, QUE J4Z1A3

SASKATCHEWAN

Fireside Conserver Products
835 Broadway
Saskatoon, SASK S7N1B5

Bibliography

Books

Adams, Margaret Byrd. *Warm and Toasty*. Seattle: Pacific Search Press, 1981.

Addkison, Andrew Roy. *Cooking On A Woodburning Stove*. Sacramento: Jalmar Press, Inc., 1980.

Annual Book of ASTM Standards #26. *Gaseous Fuels: Coal and Coke; Atmospheric Analysis*. Philadelphia: American Society for Testing and Materials, 1979.

Bartok, John W. Jr. *Heating With Coal*. Charlotte, VT: Garden Way Publishing, 1980.

Chadwick, Janet Bachand. *The Country Journal Woodburner's Cookbook*. Brattleboro, VT: Country Journal Publishing Company, Inc., 1981.

Cooper, John A., and Dorothy Malek, eds. *Residential Solid Fuels: Environmental Impacts and Solutions*. Proceedings of the 1981 International Conference on Residential Solid Fuels, Beaverton, OR, June 1–4, 1981.

Crane, Charles Edward. *Winter in Vermont*. New York: Alfred A. Knopf, 1941.

Curtis, Christopher, and Donald Post. *Be Your Own Chimney Sweep*. Charlotte, VT: Garden Way Publishing, 1979.

Darling, Dale Y., and Julia Van Dyke. *The Airtight Woodstove Cookbook*. Andover, MA: Brick House Publishing Company, Inc., 1980.

DeAngelis, D. G., and R. S. Reznik. *Source Assessment: Residential Combustion of Coal*. Dayton, OH: Monsanto Research Corporation. Prepared for the U.S. Environmental Protection Agency, January 1979.

Fryling, Glenn R., ed. *Combustion Engineering*. New York: Combustion Engineering, Inc., 1966.

Gullian, Jorden W. *Improving Your Forest for Ruffed Grouse*. Corapolis, PA: 1979.

Mining Information Services Division of McGraw-Hill Mining Publications. *Keystone Coal Industry Manual*. New York: McGraw Hill Mining Publications, 1974.

O'Connor, Hyla. *Cooking on the Woodstove*. Westport, CT: Turkey Hill Press, 1981.

Orton, Vrest. *The Forgotten Art of Building a Good Fireplace*. Dublin, NH: Yankee, Inc., 1969.

Rodale, Robert, ed. *Encyclopedia of Organic Gardening*. Emmaus, PA: Rodale Press, 1971.

Rodale, Robert, ed. *How To Grow Fruits and Vegetables By the Organic Method*. Emmaus, PA: Rodale Press, 1970.

Rose, Charles D. *Introduction to the Handling, Preparation and Use of Coal.* American Coal Training Institute. (Copyright applied for, 1979).

Schuler, Stanley, and Cary Hill. *Coal Heat.* Exton, PA: Schiffer Publishing Limited, 1980.

Segeler, C. George, ed., *Gas Engineer's Handbook.* New York: Industrial Press, Inc., 1977.

Shelton, Jay. *The Woodburners' Encyclopedia.* Waitsfield, VT: Vermont Crossroads Press, Inc., 1976.

Shelton, Jay. *Wood Heat Safety.* Charlotte, VT: Garden Way Publishing, 1979.

Shirley, Hardy L., and Paul F. Graves. *Forest Ownership for Pleasure and Profit.* Syracuse, NY: Syracuse University Press, 1967.

Stamper, Eugene, ed., and Richard L. Koral, consulting ed. *Handbook of Air Conditioning, Heating, and Ventilating.* New York: Industrial Press, Inc., 1978.

Stephens, Rockwell. *One Man's Forest.* Brattleboro, VT: The Stephen Greene Press, 1974.

Twitchell, Mary. *Wood Energy.* Charlotte, VT: Garden Way Publishing, 1978.

Vivian, John. *Wood Heat.* Emmaus, PA: Rodale Press, 1976.

Pamphlets and Periodicals

American Forest Institute. *Forest Facts and Figures.* Washington, D.C.: 1978.

Anthony, Tony. "Chimneys," *Wood 'N Energy* (March, 1982), p. 25.

Hewitt, Charles E. et al. "Wood Energy in the United States," *American Review of Energy,* 1981.

Hewitt, Charles E., and Colin J. High. "Environmental Aspects of Wood Energy Conversion." Paper for U.S. Department of Energy Environmental Control Symposium, Washington, D.C., November 28–30, 1978.

Lancaster, Ken, and L. Hunt. *Improve Your Woodlot By Cutting Firewood.* Upper Darby, PA: USDA Forest Service, State and Private Forestry.

Myer, Albert J., ed., *Woodstove Directory.* Manchester, NH: Energy Communications Press, Inc., 1982.

Olson, D., and Clarence Langer. *Care of Wild Apple Trees.* USDA Forest Service, N.E. Area, NAFB/M-S.

National Coal Association. *Coal Facts.* Washington, D.C.

National Fire Protection Association. *Using Coal and Wood Stoves Safely,* No. HS-10. Boston: 1978.

State of Vermont Agency of Environmental Conservation. *Landowners Guide to Wildlife Habitat Management for Vermont Woodlands.*

State of Vermont Agency of Environmental Conservation. *Guide for Controlling Soil Erosion and Water Pollution on Logging Jobs in Vermont.*

Stevens, Daniel, and Lois Frey. *Buying Firewood. Morrisville, VT: Northern Vermont R.C. and D. Area.*

Index

About the Authors

Stephen Morris has been at various times a rock musician, a magazine editor, and a freelance writer. As Manager of Customer Relations his face and signature are familiar to the many Vermont Castings' customers who read the *Owners' News*.

A self-proclaimed beer expert, he has pioneered the art of brewing on the wood stove. He is the founder of the Cram Hill Brewers and claims the distinction of having visited every independent American brewery. His beer belly is kept at a minimal level only by the many miles he logs running and cross-country skiing in the hills surrounding West Brookfield, Vermont where he lives with his wife and two sons. He is a lifelong New Englander and a graduate of Yale College.

Vance R. Smith came to Vermont by way of Southern California, Wellesley College, and Harvard University where she received her master's degree in architecture. A prominent local illustrator lured to Vermont by her interest in alternative energy, her work has appeared in *Yankee* and *Country Journal*, as well as being closely identified with the Vermont Castings' operation manuals, and their promotional and technical literature. She counts *The Woodburners' Encyclopedia* among her previous illustrating credits.

Vance plays first base for the Vermont Castings' softball team. She raises turkeys, and her lettuce is among the earliest in Central Vermont. She is an enthusiastic supporter and on the board of directors for the A. B. Chandler Cultural Foundation in Randolph.

Bill Busha grew up in western Massachusetts near the Connecticut River. His childhood home was a large and drafty house heated by a coal and wood burning furnace, and to this day he retains characteristics developed during that period of his life: He dresses with lightning speed each morning and seldom returns from a leisurely walk without an armful of kindling. He graduated from American International College and soon after was drafted by the Army to spend the next two years in Tokyo, Japan.

Upon his return home he settled in Vermont and taught for several years at a local high school. He received a graduate degree from the University of Vermont, and his interest in wood heat eventually led to a position at Vermont Castings. Bill is the senior member of the Customer Relations department and the editor of the *Owners' News*.

Bill lives with his wife and twin sons in Randolph, where he is a trustee of the public library. Many of his after work hours are spent restoring a Vermont hillside to productive use. He manages a small flock of sheep, a large garden, and a modest sugaring operation.

NOTES

NOTES

NOTES

NOTES